CREATION *as* SCIENCE

CREATION
as SCIENCE

A TESTABLE MODEL APPROACH TO END THE CREATION/EVOLUTION WARS

HUGH ROSS

NAVPRESS®

BRINGING TRUTH TO LIFE

OUR GUARANTEE TO YOU

We believe so strongly in the message of our books that we are making this quality guarantee to you. If for any reason you are disappointed with the content of this book, return the title page to us with your name and address and we will refund to you the list price of the book. To help us serve you better, please briefly describe why you were disappointed. Mail your refund request to: NavPress, P.O. Box 35002, Colorado Springs, CO 80935.

The Navigators is an international Christian organization. Our mission is to advance the gospel of Jesus and His kingdom into the nations through spiritual generations of laborers living and discipling among the lost. We see a vital movement of the gospel, fueled by prevailing prayer, flowing freely through relational networks and out into the nations where workers for the kingdom are next door to everywhere.

NavPress is the publishing ministry of The Navigators. The mission of NavPress is to reach, disciple, and equip people to know Christ and make Him known by publishing life-related materials that are biblically rooted and culturally relevant. Our vision is to stimulate spiritual transformation through every product we publish.

ISBN 1-57683-578-2

Cover design by Jonathan Price
Cover illustration by Stock Illustration Source/Getty Images
Creative Team: Terry Behimer, Andrea Christian, Linda Vixie, Kathy Mosier, Arvid Wallen, Pat Reinheimer

Some of the anecdotal illustrations in this book are true to life and are included with the permission of the persons involved. All other illustrations are composites of real situations, and any resemblance to people living or dead is coincidental.

Unless otherwise identified, all Scripture quotations in this publication are taken from the HOLY BIBLE: NEW INTERNATIONAL VERSION® (NIV®). Copyright © 1973, 1978, 1984 by International Bible Society. Used by permission of Zondervan Publishing House. All rights reserved.

Published in association with the literary agency of Alive Communications, Inc., 7680 Goddard Street, Suite 200, Colorado Springs, CO 80920 (www.alivecommunications.com).

Ross, Hugh (Hugh Norman), 1945-
  Creation as science : a testable model approach to end the
creation/evolution wars / Hugh Ross.
     p. cm.
  Includes bibliographical references and index.
  ISBN 1-57683-578-2
  1. Religion and science. 2. Creation. 3. Cosmogony. I. Title.
  BL240.3.R67 2006
  231.7'652--dc22
                         2006013611

Printed in the United States of America

1 2 3 4 5 6 / 10 09 08 07 06

FOR A FREE CATALOG OF NAVPRESS BOOKS & BIBLE STUDIES,
CALL 1-800-366-7788 (USA) OR 1-800-839-4769.

For all the Reasons To Believe volunteer
apologists and chapter members

CONTENTS

FIGURES AND TABLES

Figures

Tables

ACKNOWLEDGMENTS

Each of my books has represented a team effort, but none more than this one. I'm indebted to my colleagues Fuz Rana, Dave Rogstad, Kenneth Samples, and Jeff Zweerink for their contributions to the development and articulation of the Reasons To Believe (RTB) creation model, which this book presents. Their research and insights have been invaluable to me for years and certainly gave shape to the material you're about to read. Marj Harman deserves credit as the most patient and meticulous research assistant anyone could wish for.

Given that books often *are* judged by their covers, at least initially and despite advice to the contrary, the fact that you're reading these words may be due to the talents of Jonathan Price, RTB's art director, who applied his skill to the cover design as well as to the various explanatory figures throughout the book.

Thanks to Diana Carree and Sandra Dimas for help with all the pesky but vital details I tend to overlook. Only through their focus and efficiency were deadlines met.

Much credit for the organization and readability of the pages to follow goes to my dedicated editorial team, led by Kathy (my wife and "translator") and Patti Townley-Covert. Other significant editorial guidance came from Rachelle Gardner, Tani Trost, Andrea Christian, and Linda Vixie.

Special thanks to Dan Rich and Terry Behimer of NavPress and to Lee Hough of Alive Communications for encouraging me to tackle this project, which they and I dared to envision as a valuable resource for both skeptics and believers amid the escalating—and very public—controversy over creation, evolution, and intelligent design.

I wish I could name all the people, including university science faculty and RTB volunteer apologists, whose comments and questions in response to my lectures over the past few years have influenced my thinking and writing. I appreciate their feedback more than words can express.

Hugh Ross

SIFTING FACT FROM FICTION

F or decades after its publication in 1898, H. G. Wells' classic novel, *The War of the Worlds*, captivated readers. Though Wells' fictional account of a Martian invasion was widely familiar, the 1938 radio dramatization by Orson Welles came across with such realism that it generated widespread panic.

In Pittsburgh, a man came home to find his wife with a bottle of poison in her hand—convinced that death was the only way to avoid capture by the diabolical aliens. A hospital in New Jersey treated more than a dozen men and women for shock and hysteria.[1] In Hollywood, actor John Barrymore let loose his 10 beloved Great Danes to fend for themselves because, in his words, "the world is finished."[2] Across America, thousands of people acted according to their belief that human life on planet Earth was about to end.

Several factors played into the broadcast's believability. In an atmosphere of growing uneasiness over the threat of war in Europe, the public had become accustomed to news bulletins regularly interrupting radio broadcasts. Welles lent credibility to the supposed invasion by using the news bulletin format, with authoritative voices realistically portraying a Princeton professor and various military officials. Just as belief impacts action, emotion impacts belief. And the higher the stakes, the more intense the emotion may be.

THE TWO HUNDRED YEARS' WAR

During the 17th century, 10 European powers engaged in a protracted, bloody war over religion and politics. It went on for so many years that eventually its duration became its identity: the Thirty Years' War. By comparison, the current conflict· between creation and evolution—a clash of worldviews, methodology, and politics—has raged for more than two centuries. Proponents on all sides of the issue—atheists, agnostics, pantheists, deists, and theists—slash each other with saber tongues as they debate the origin and unfolding history of the universe and of Earth, of life in general, and of humanity in particular. Their marshaled "facts" seem to collide head-on.

- Astronomers say the universe is 13.73 billion years old[3] and Earth, 4.5662 billion years old.[4]
- Many Christians, as well as some Jews and Muslims, say the universe and Earth have been around for only a few thousand years.
- NASA anticipates the discovery of indigenous extraterrestrial life.
- Some scientists contend that life's manifold finely tuned requirements make such a search ludicrous.
- News reports announce discoveries of "missing links" between early primates and humans.
- The Bible says God created Adam and Eve.

What's really going on? How can a person know what to believe? Is there any way to sort out facts from fables?

ABOUT THIS BOOK

In responding to *The War of the Worlds* broadcast, listeners had to determine whether the Martian invasion was really happening or not. The war between the realms of creation and evolution involves discerning reality as well. Aristotle reminds us, with his timeless logic, "that everything must be either affirmed or denied, and that a thing cannot at the same time be and not be."[5] Either everything that exists, including life and humanity, is the natural outcome of a self-existent universe, or it's not. Either the universe was designed with purpose and meaning, or it was not. The entire cosmos either explains itself, or it does not. Creation either happened, or it did not.

Unfortunately, discerning what's true about the origin and history of the universe, Earth, life, and humanity remains difficult—not just because the science is challenging, but also because of ongoing battles between those most involved in the search for answers: people of science and people of faith. This book begins by addressing the unrelenting antagonism among advocates of various views. As a scientist and a pastor I've been caught in the crossfire on university campuses and in business settings, churches, and seminaries. Evolutionists, creationists, and ID-ologists (see "Who Are the Combatants?" pages 20–21) continue to blast away at each other.

I believe an alternative to this type of battling does exist, and I am not alone in this view. Many science researchers, Sunday school teachers, preachers, professors, professionals, and laypeople agonize over the hostilities and believe a way can *and must* be found to pursue answers with civility—without compromising integrity.

As a scientist and the founding president of Reasons To Believe (RTB), I have been involved for many years in an effort to minimize the raging antagonism, encourage dialogue, and develop credible interpretations of the data. With help and encouragement from many researchers, RTB has been engaged in the process of developing a scientifically testable model (see "What Is a Model?" page 32) to investigate and explain the origin and history of the universe, Earth, life, and humankind. The foundations of this model, the RTB creation model, are set forth in chapter 4 and further elaborated in the subsequent five chapters.

Much as listeners of *The War of the Worlds* had to test various aspects of the broadcast to assess its credibility, inquiring minds must analyze the measured characteristics of the universe and life to determine how well they align with a variety of explanations—various evolutionary theories, the biblical creation scenario, and all other creation perspectives.

The diverse positions on evolution and creation really can be placed in the realm of testability. First, investigators ask probing questions and collect all the available (relevant) data. Then they can review or revise various explanations, develop and apply critical tests, and make logical predictions, in light of their models, about what each model anticipates researchers should discover in their ongoing studies. Because this model approach has already been demonstrated to some degree in previously published RTB books (*The Creator and the Cosmos, Origins of Life, Who Was Adam?, A Matter of Days,* and *Lights in the Sky and Little Green Men*[6]), this volume sets forth the broader context and use of the RTB creation model.

Within the limits of a single book, only the most significant features of the RTB model (at its current state of development) can be presented. I have attempted to show (1) how the RTB creation model can be either falsified or affirmed, and (2) how well it has fared to date in explaining the record of nature and predicting scientific discoveries. In order for readers to assess how well the model succeeds, compared to other creation/ evolution models, in anticipating new discoveries, this volume includes the RTB creation model's forecasts—what we expect researchers to find during the next several years if the model is reasonably accurate.

My own search for answers to creation and evolution questions has yielded stunning results, dramatically impacting my views and beliefs. I see uncanny and uncompromising harmony between the facts of nature and the words of Scripture. New research and discoveries continually stir the passion that fuels my work. They also supply a never-ending motivation to refine my views. These insights, based on science and the Bible, have led to the development of RTB's creation model and to the content of this book:

- Chapters 1 and 2 describe the most publicized creation and evolution positions, the strategies and principles behind them, and the legal concerns of bringing religion/theology into science research and education.
- Chapter 3 sets forth a method for testing reality and compares the explanatory success of various models for the origin and history of the universe, Earth, life, and humanity.
- Chapter 4 develops the biblical foundations for the RTB creation model and identifies many of its scientifically testable features.
- Chapters 5–9 examine and extend the RTB creation model in light of the latest scientific findings about the origin and development of the universe, solar system, life, and the human species. (Readers who have little or no interest in the scientific details may choose to skip these chapters and go on to chapter 10.)
- Chapters 10–12 look to the future to anticipate how the RTB creation model and other creation/evolution models will be strengthened or weakened by further scientific discoveries and to explore how barriers between the various creation/evolution camps can be removed or at least diminished. The final chapter shows how freedom from censorship can make science education captivating once again. The goal is to facilitate a more productive pursuit of truth and engage the public in increasingly enthusiastic support of both scientific and theological enterprises.

Part I

THE
PERENNIAL
CONFLICT

CHAPTER 1

THE CREATION/EVOLUTION BATTLEFIELD

When Orson Welles' 1938 radio dramatization of *The War of the Worlds* threw thousands of people into panic—stirring them to say frantic good-byes, flee their homes and cities, and even prepare to take their own lives—pundits roared at the public's gullibility. It's tempting to nod in agreement with the *New York Tribune* columnist who praised both Welles and CBS for making an "important contribution to the social sciences" by casting "a brilliant and cruel light upon the failure of popular education."[1] In the words of one nine-year-old boy, the people who believed the invasion was real must have been "dumb-bells."[2]

Readers who have listened to the broadcast or read its transcript may be more charitable in their judgments. Given Welles' theatrical brilliance, his plan to pull off a Halloween prank of national proportions was almost certain to succeed. Asked later about his motivation for creating such a realistic dramatization, Welles explained that he "wanted people to understand that they shouldn't take any opinion predigested, and they shouldn't swallow everything that came through the tap, whether it was radio or not."[3]

Radio, television, books, newspapers, the Internet, and the person in the next cubicle all supply "facts" and opinions that influence one's beliefs. In the case of the 1938 broadcast, an undercurrent of fear and dread, along with preconceptions and limited information, contributed to the ensuing pandemonium. Some of these same factors heighten the conflict over creation and evolution issues today, instigating one battle after another. For a glimpse at how caustic and personal these battles have become, see appendix A, "Creation/Evolution Verbal Warfare," pages 205–207.

WHO ARE THE COMBATANTS?

Creationists

Historically the term "creationist" has applied to anyone who acknowledges that a Creator is responsible for bringing the universe and life into existence. In that sense, it pertains to nearly half of all practicing scientists (see chapter 2, pages 27–28).[4] Over the past several decades, however, in the face of emerging scientific challenges to traditional beliefs, the term has taken on a much narrower meaning. Today "creationist" typically refers to someone who believes that:

- The Genesis creation days must be six consecutive 24-hour periods, and God created all things in that span.
- The Genesis genealogies contain few if any gaps, and therefore, the creation week occurred between 6,000 and 10,000 years ago.
- The flood of Noah's time (Genesis 6–9) was a global event, submerging all the continents and destroying all land-dwelling, air-breathing animals (except those aboard the ark).
- All land animals on Earth today naturally descended from (at most) the few tens of thousands of pairs of creatures on Noah's boat.
- The flood of Noah's day accounts for virtually all the geological features, fossils, and biodeposits found in Earth's crust.

Several parachurch organizations have advanced this set of teachings, commonly referred to as "creationism" and "creation science," so effectively that many evangelical pastors, congregations, schools, broadcasters, ministry leaders, and missionaries adhere to them more or less by default and remain largely unaware (or distrustful) of any alternative biblical view of creation. Most reporters, pundits, and secular scientists see these teachings as part of all evangelicals' belief system. Thus, the terms "creationism," "creationist," and "creation science" as used in this book refer to that common meaning.

However, the word "creationist" may occasionally be used with a qualifier to refer to someone who, by contrast, believes the biblical account of creation *and* gives credence to the findings of science. These individuals typically embrace both the truthfulness of Scripture and the scientific evidence for a multi-billion-year history of the universe, Earth, and life on Earth.[5]

Evolutionists
Scientists initially used the term "evolution" with reference to nature's change over time—change brought about by any and all means. By this broad definition, even the Bible describes evolution. In recent decades, however, the word "evolutionist" has generally been applied to someone who asserts that all the changes observed in the record of nature (including the origin and history of the universe, Earth, and all life) can be attributed to strictly natural causes. Unless otherwise qualified, the terms "evolutionist," "evolutionism," and "evolution science" in this book refer to this core belief.

Intelligent Design Movement
For thousands of years, scholars from various cultural and religious backgrounds have proposed intelligent design as an explanation for many of the special properties of the universe, Earth, life, and humanity. For over a century, every student at Britain's Cambridge University was required to study William Paley's famous text *Natural Theology*, in which Paley infers from his detailed study of nature that the properties of living organisms demand a divine Creator.

Even apart from questions about how the universe and life began, intelligent design has long been acknowledged as a legitimate scientific conclusion. In such disciplines as archeology, anthropology, and forensics, researchers evaluate, differentiate, and interpret evidence or artifacts based on various indicators of intentionality or purposeful design.

About a decade ago, however, a diverse group of creation advocates formed an alliance that has become widely known as the Intelligent Design movement (IDM) (see chapter 2, pages 30–33). Their goal is to advance public instruction of the intelligent design concept, the inference that an intelligent designer is responsible for the origin and history of life. By refraining from making a specific identification of the designer, the movement has sought to remove any particular religious bias and therefore any apparent legal basis for disallowing the teaching of intelligent design in America's classrooms.

WHY ARE THEY FIGHTING?

Each of the major combatants in the protracted creation/evolution battle wants exclusive rights to the story of the cosmos and of life—a story that carries enormous significance for every person on Earth, past, present, and future. That story holds the key to unlocking the great mysteries of life: Why (and whence) did humanity get here? Where is humanity headed?

Did God (or gods) spring from human imagination, or did human imagination come from God? Who or what determines the meaning of life?

Winning the position of authority on that story seems to have eclipsed all other objectives. At least that is what the strategies and tactics of the combatants suggest (see chapter 2). Is there any basis for hope, then, that any of the creation/evolution controversies will be resolved?

Delos McKown, chairman of the philosophy department at Auburn University, wrote: "The twenty-first century will likely witness, as never before, the battle of science with soteriology [the doctrine of salvation], of free enquiry with religious dogma desperately held and tenaciously maintained."[6] He bases this dismal forecast of ongoing conflicts on an evident trend: despite amazing advances in research, the creation/evolution controversies proceed not toward resolution, but rather toward greater polarization and entrenchment. Such a clear and persistent direction leaves little room for optimism.

McKown's words supply an important clue, however, both to the problem and to its solution. He refers to "religious dogma" (presumably creation doctrine) and contrasts it with "free enquiry." His phraseology implies that the Christian approach to creation is rigidly fixed and absolute, neither proved nor provable. However strongly Christians may object, this characterization is widely accepted and largely true, not just in America but worldwide.

The diversity of opinion within Christianity—even among evangelicals—on such issues as predestination and free will, the nature of the sacraments, the essentials of salvation, and the future of the world divides Christians into many denominations. Yet the world still recognizes this panoply as essentially Christian, and some Christians consider this diversity positive and healthy. Why, then, is enquiry on the topic of creation considered off-limits by so many Christians and non-Christians alike?

One contributor to the "closed-minded" Christian image is the sad fact that many Christian churches, schools, broadcasters, and book/audiovisual distributors have shown themselves unwilling to consider—or even to hear—alternative interpretations of the biblical creation texts.[7] Australian geologist Ian Plimer observes, "Criticism of creationism is just not tolerated; it is avoided at all costs and every effort is made to silence, discredit or belittle the critic."[8] To make matters worse, the familiar young-earth view makes Christians so easy to marginalize and ridicule that mainstream media and academia often reinforce this view as *the* Christian position and capitalize on it.[9]

Another contributor to the problem is that creationism's high profile silences many scientists and other scholars of faith from mentioning their creation beliefs in public. They keep quiet, not wanting to be branded as "evolution-phobic" or as skeptics of mainstream science who advocate placing religious dogma into public school curricula.[10] Consequently, the public rarely, if ever, hears about (or from) Christian scholars (not to mention other critics of Darwinism) and their variety of views on creation.

A CALL FOR GREATER TOLERANCE *AND* DISCRIMINATION

While discrimination against people must be avoided at all costs, failure to discriminate between truth and fiction represents a serious problem. All *people* are created equal, but not all *truth-claims* are equally valid. Tolerance among people who hold different beliefs about origins reflects an appropriate respect for diversity. However, not all beliefs or interpretations correspond equally well with physical reality. Some correspond well, some correspond poorly, and others flatly contradict.

Philosopher J. P. Moreland says, "Reality makes propositions true or false. A proposition is not made true by someone's thinking or expressing it, and it is not made true by our ability to determine that it is true. Put differently, evidence allows us [to] tell if a proposition is true or false, but reality (the way the world is) is what *makes* a proposition true or false"[11] (emphasis in original). In *The War of the Worlds*, a long-legged water tower *resembled* a Wellesian Martian. In the dark of night, a man *believed* the tower to be a Martian. However, the structure he shot was still just a water tower. Daylight revealed the cold, hard (bullet-scratched) reality.

Exploring the physical evidence related to the origin and history of the universe, life, and humanity can illuminate the strengths and limitations of a particular interpretation. Asking appropriate questions can help reveal which explanation most closely corresponds to facts. Does the Darwinian approach offer an adequate explanation? Is a conclusion of a recently created Earth and universe warranted either scientifically or biblically? Do human beings possess unique spiritual properties? Just as importantly—can a particular position withstand the testing afforded by new discoveries and new understanding? All credible evidence must be considered when determining how closely a particular explanation (or prediction) corresponds to reality.

In this healthy competition of ideas, tolerance and discrimination can and must be exercised in balance. Christian scholars ask their peers in the science community to demonstrate balance by a willingness to entertain the possibility of both natural and supernatural explanations for life's origins and development. Theoreticians have already demonstrated that willingness in the arena of cosmology, the study of the origin, structure, and development of the universe (see chapter 3, pages 44–51). Meanwhile, the science community challenges Christians to demonstrate some balance among themselves by discriminating against interpretations and explanations that contradict established data.

Discrimination is not to be confused with censorship. Major participants in the creation/evolution conflict have attempted to stonewall discussion of each other's models. Thus, they have barred each other's models and, consequently, have forfeited the benefits of evaluation, critique, and testing. Such censorship benefits no one (see chapter 12, pages 195–196).

For healthy, productive discrimination to occur, more creation/evolution models must be put on the table, models that can be verified or falsified, models that effectively predict future scientific discoveries. This book highlights one such model (still in progress), the Reasons To Believe (RTB) creation model, emphasizing the model's testable and predictive features. This model is offered in the hope of providing a viable, broadly integrative explanation—superior to previously proffered explanations—for the universe's and life's origins and histories. It is offered in the hope of encouraging development of competing models more testable and predictively successful than currently existing ones.

The challenge to consider all credible positions represents an exhilarating opportunity. It offers a way for all creation and evolution debate and research participants to regain some measure of respect and credibility. By replacing fear, preconceived ideas, superficial knowledge, and a lack of sophistication with rigorous scientific standards, set and maintained along the lines of Moreland's definition of truth, this quest could yield unprecedented breakthroughs. But first, unpacking the strategies and animosities that separate the most widely publicized positions in the creation and evolution conflict may lead toward the kind of understanding, the *cease-fire*, necessary for objective testing to begin.

CHAPTER 2

BATTLE PLANS:
WHO'S SAYING WHAT?

Careful attention to the opening of *The War of the Worlds* broadcast might have prevented the listener panic of 1938. The announcer clearly introduced the program as a theatrical presentation. Those who had read the book by H. G. Wells should have recognized the story line. If listeners had paid attention to the fact that a commercial ran during the show, they might have been entertained rather than terrified.

Those who tuned in a few moments late were especially prone to believe that Martian invaders were fanning out from New Jersey to overrun the world. Their immediate emotional reactions prevented them from responding rationally. Perhaps the intensity in the actors' voices pulled them away from their anchors to reality.

The topic of creation and evolution carries people along similarly emotional pathways. Because so much is at stake, the spokespersons for various positions tend to communicate with passion and hyperbole. A listener (or reader) must work hard to resist being swept away. This chapter examines the most visible strategies creationists and evolutionists employ in their efforts to win the high-stakes worldview war. It evaluates the effectiveness of those strategies and points toward possible breakthroughs.

Seventy-five years after the Scopes Trial (see appendix B, "Does the Constitution Bar Creation Teaching?" pages 209–218) brought the creation/

evolution conflict into the national—even global—spotlight, renowned evolutionist Niles Eldredge, in his book *The Triumph of Evolution and the Failure of Creationism*, boldly declared evolution's victory.[1] What victory? Although evolution has gained and retained prominence in public education, it has failed to vanquish belief in creation.

The question is, why? Are people simply misinformed? What keeps belief in creation alive?

CREATION'S GRIP

A 1916 survey of 1,000 randomly selected scientists assessed certain aspects of their religious beliefs. Psychologist James Leuba found that 40 percent believed in a personal God and an afterlife, 40 percent disbelieved, and 20 percent were unsure.[2] When he compared these numbers with statistics for the population as a whole, of which 90-some percent believed in God, Leuba concluded that scientific literacy accounted for the difference. Consequently, he predicted that with the growing breadth and depth of science education, belief in God would decline. Another prediction claimed that with the advance of scientific discovery, the percentage of scientists believing in God would rapidly and severely dwindle.

In the early 1900s, few people completed high school and only a small fraction had access to college education. Today, a much larger percentage of students finish high school and pursue at least some college instruction. Science classes are standard requirements. The body of data that informs these courses has multiplied enormously since 1916. People of the 21st century know far more science than their predecessors.

Yet statistics confirm that roughly 90 percent of Americans still believe in God,[3] and 83 percent believe that God played the primary role in creating and shaping life.[4]

A recent Harris Poll shows that belief in the familiar evolutionary explanation for human origins has actually waned somewhat over the past decade (see table 2.1, below).

TABLE 2.1:
HARRIS POLL RESULTS ON WHETHER HUMANS WERE CREATED BY GOD

A nationwide poll of 1,000 U.S. adults surveyed by telephone by Harris Interactive between June 17 and 21, 2005, produced the following results:[5]

QUESTION	ANSWER	MARCH '94	JUNE '05
Do you think human	Yes	44%	38%
beings developed from	No	46%	54%
earlier species or not?	Not sure	11%	8%
Do you believe apes and	Yes	51%	46%
man have a common	No	43%	47%
ancestry or not?	Not sure	5%	7%

The question "Which of the following do you believe about how human beings came to be?" yielded the following responses:

Human beings evolved from earlier species	22%
Human beings were created directly by God	64%
Human beings are so complex that they required a powerful force or intelligent being to help create them	10%
Not sure	4%

This finding that most people still doubt naturalistic explanations for life, including humanity, has prompted comments such as this one from columnist Peter Schrag:

> Given the polls showing that most Americans don't accept Darwinian evolution and know little about it . . . if creationists and intelligent designers really believed in equal time, they'd demand more emphasis on evolution, which doesn't seem to be getting equal time in most Americans' heads.[6]

Even professional scientists reflect little change in their views over the past century. A 1996 survey of 1,000 scientists randomly selected from the latest edition of *American Men and Women of Science* proved Leuba's 1916 prediction incorrect. Of those questioned, 40 percent indicated their belief in a personal God and an afterlife, 45 percent said they disbelieve, and 15 percent identified themselves as agnostic.[7] Neither laypeople nor scientists appear to have been significantly influenced, at least with respect to belief in God, by education's emphasis on evolution.

So What's Going On?

The statistics baffle both evolutionists and creationists. More questions are raised than answered. Evolutionists must ask why, given their steady grip on public education and the remarkable advances in science, naturalistic teaching fails to impact people's beliefs. Evolutionary biologist Edward O. Wilson remarks in his newly published anthology of Charles Darwin's principal writings that public disbelief in evolution is "surpassingly strange."[8]

Young-earth creationists must ask why—given the persistence and prevalence of belief in creation and the pervasiveness of their creation teaching—their views have failed to impact what's taught in public schools and colleges,[9] even in many (or most) evangelical colleges and seminaries.[10]

Two reasons seem obvious, one on each side.

An insistence on strict naturalism thwarts the evolutionist cause. If nature (the observable, physical universe) is the only reality, then all mysteries are simply material episodes, all thoughts and feelings simply the result of physical interactions, and all events take place without meaning beyond mere sequential timing. The big questions of life become insignificant. *Why are people here?* We just are. *Where did humans come from?* A random physical phenomenon. *Where are they going?* Most likely, to extinction unless people rapidly evolve into a different life-form or find a new home.

Such answers rarely captivate a student's curiosity or ignite imagination. And they don't satisfy the soul's hunger for meaning, purpose, or hope.

On the other hand, the insistence on a particular, subjective interpretation of the biblical creation account thwarts the young-earth creationist enterprise. Focusing primarily on Genesis and overlooking the many other relevant Bible passages on creation, these people conclude that virtually all of science is either wrong or illusory. Scientists are either deluded or conspirators. Thus, young-earth creationists place their ideas above evaluation by the scientific community, and in so doing they make no impact on the scholars who shape school curriculum. They seem to have forgotten that curriculum changes follow, rather than precede, paradigm shifts.

Legislative Battles Rage

Despite academia's resistance, young-earth creationists have worked hard since 1925 to make inroads to public instruction through grassroots political action. They have lobbied local school boards and state legislatures for "equal access" (equal time for young-earth views) whenever and wherever evolution is publicly taught.

Though occasionally equal access laws are passed, none has yet gone into effect. Various organizations and citizens groups have consistently challenged these rulings in court, where state and federal judges have overturned them. The United States courts have repeatedly blocked attempts to mandate young-earth teaching in a public context.

Leading creationist proponents usually claim that judges and justices are motivated by an anti-Christian bias and a misguided application of the U.S. Constitution. A reading of the case documents can easily remove doubts about the basis for those rulings: the courts have ruled against them because their cases lacked scientific merit. (See appendix B, "Does the Constitution Bar Creation Teaching?" pages 209–218.)

In particular, the United States Supreme Court justices have pointed out that "requiring the teaching of creation science with evolution does not give schoolteachers a flexibility that they did not already possess to supplant the present science curriculum with the presentation of theories, besides evolution, about the origin of life."[11]

The Supreme Court justices affirmed that if creationism was valid as science, its right to a place in the public school science curriculum would be assured—no matter what its connection to religion.[12] Based on testimony and documentation from both Christian and non-Christian experts, however, various courts have found a complete lack of scientific credibility and integrity with this young-earth position. So, even if the proposed creationist instruction had no connection whatsoever with religion, it would still be ruled out. The appeal to protected equal access for public school science curriculum was appropriately denied.

According to judgments in all the major U.S. cases over the past 50 years, scientific credibility was the issue at stake. The courts acknowledged that neither scientific perfection nor completeness is required, but a reasonable level of plausibility and integrity must set the standard. In the courts' assessment, young-earth creationism falls short of the minimal level.

These judgments are not bold judicial maneuvers. They are, in fact, historically consistent decisions. For example, until the 1920s a sizeable group of U.S. citizens believed (according to their "literal" reading of the Bible) that Earth was flat rather than spherical.[13] Though Christian proponents of this view were serious and devout, their religious belief did not justify inclusion of flat-Earth science in school curriculum. The same curriculum standard holds true for geocentrism (belief that the earth rather than the sun is the center of the solar system), part of the religious landscape for the past 250 years but still not taught in school.[14]

Opening Up a Different Front

Repeated rulings against creationism made the necessity of an alternative approach obvious. Most recently the Intelligent Design movement (IDM) has tried to reclaim the creationist's place in public education under the banner of religious neutrality. Based on the assumption that courts ruled against young-earth creationism primarily due to its religious agenda, Intelligent Design (ID) leaders assembled a diverse coalition of non-naturalist scholars and organizations to strip all visible signs of religious affiliation from creation teaching.

In a series of four books, ID spokesman Phillip Johnson laid out his plan for restoring the damaged image and influence of creation advocates.[15] He referred to building the coalition as a "big tent" and to its mobilization as driving in a "wedge of truth." Johnson's strategy reflects his expertise and experience as a legal scholar.

The big tent includes deists, atheists, agnostics, Mormons, Muslims, Jews, Roman Catholics, Eastern Orthodox, Protestants, evangelicals, and even some extraterrestrial intelligence advocates who believe life arrived via aliens. Such a polyglot of religious perspectives would seem to prevent any challenge based on the First Amendment of the U.S. Constitution: "Congress shall make no law respecting an establishment of religion, or prohibiting the free exercise thereof."

Johnson also argues that only through such a broad-based coalition is there any hope of mobilizing the necessary talent and financial resources to challenge and overthrow creationists' mighty foes: naturalism and the evolution establishment. He calls the big tent a "unite and win strategy"[16] and strongly urges all anti-naturalists to put aside their different creation perspectives and disputes with one another until their common foes are utterly vanquished.

Many ID theorists suggest that the easiest creation truth to establish in secular circles is the existence of some type of intelligent design (as opposed to mindless Darwinism) behind the record of nature. Once that truth becomes firmly established in secular society, Johnson and his colleagues are persuaded that other basic creation truths could be built on this basic ID concept, including truths about the kind of divine Being responsible for intelligent design.

Scientists' reaction to intelligent design advocacy has been dismissive and hostile, however. Robert Pennock, philosophy of science professor at Michigan State University, describes certain ID leaders as people "who want not just absolute truth, but their unjustified, anti-scientific version of it."[17] Evolutionary biologist Jerry Coyne refers to the research and writing from ID scholars as "stealth creationism."[18] Physicist Mark Perakh accuses ID leaders of "disdainful dismissal of all and every criticism."[19]

Lawrence Krauss, director of the Center for Education and Research in Cosmology and Astrophysics (CERCA) at Case Western Reserve University, says ID definitely is not science.[20] Jennifer Palonus, director of the online forum *Creation/Evolution: The Eternal Debate*, says intelligent design proponents "have not been able to articulate a positive scientific case for 'intelligent design.' ID is still just a collection of negative claims about evolution."[21] (See "Evolution Bashing," below.)

While such negativity toward ID proponents may seem harsh and bigoted, it makes sense to many scientists. Reaction to design advocacy arises partly from its association with young-earth creationism. Scientists have noticed the fact that significant financial backing for the IDM's scholars and spokespersons comes from young-earth creationists, who also buy the majority of ID products.[22] Evolutionary science researchers are suspicious of ID's reluctance to offend or debate what they consider unscientific or antiscientific beliefs.

Evolution Bashing

Among scientists and reporters, creationists and intelligent design proponents have earned the label "evolution bashers." Arguments for creation tend to focus strictly on negatives. Many Christians can point out the problems and shortcomings of evolutionary theory, but few can offer a definitive explanation or evidential defense of their own beliefs about life's origin and history.

Evolutionists interpret this bashing as an act of veiled cowardice. They say if creation proponents really had the courage of their convictions, they would say what they believe — in enough detail to invite meaningful critique. For example, Philip Kitcher, a philosophy of science professor at Columbia University, asked: "What's the nonevolutionary explanation? Johnson doesn't tell us....Johnson's attempt to dispute the 'fact of evolution' is an exercise in evasion."[23]

The evolutionists' point is well taken: it's easy to sit on the sidelines and snipe at others' problems and shortcomings, but it's tough (as well as a mark of valor and integrity) to expose one's own beliefs and interpretations to public scrutiny and criticism.

Scientists repeatedly ask ID leaders at public meetings to clarify their stance on young-earth creationism. A persistent unwillingness to give a clear answer to this question reinforces their notion that ID may simply represent a stealth creationist movement. Astrophysicist Adrian Melott's headline in *Physics Today* claims "Intelligent Design Is Creationism in a Cheap Tuxedo."[24]

Carl Wieland, himself a young-earth creationist, pinpoints another major reason for such a negative response. He notes that ID theorists "refuse to be drawn in on the sequence of events, or the exact history of life on Earth, or its duration."[25] Therefore, they "can never offer a 'story of the past.'"[26] He asks, "If the origins debate is not about a 'story of the past,' what is it about?"[27]

When it comes to the origin of the universe, life, and humanity, scientists want history's story. They emphatically request that the story be cast in the form of a testable model (see "What Is a Model?" below). In a two-hour prime-time national television debate in 1997 between evolutionist and ID leaders, the evolutionists repeatedly asked, "Where is your model?" and never received a reply.[28] Nine years later, evolutionists still ask the same question and still receive no response.[29]

What Is a Model?

In science, the term "model" refers to the schematic description of a system (or set of phenomena) that accounts for its observed and inferred features. A model is much more than a mere idea, inference, method, hypothesis, or rudimentary theory. It's a scenario that offers reasonable explanations for the entire scope or history (start to finish) of a particular system or set of phenomena in nature, as well as for its relationship to other phenomena.

Use of a model approach gives researchers enough detail to assist them in further study. It offers explanations for how, when, where, in what order, and why a phenomenon takes place. It anticipates discoveries that could either verify or falsify its explanations. The best models also yield specific suggestions for near-future research studies that may help improve understanding of the systems or phenomena they attempt to explain.[30]

Without a model, the ID paradigm cannot be tested or falsified, nor can it generate significant predictions of future scientific discoveries. This lack of substantive testability gives rise to the repeated charge that ID is "not science."

To consider an alternative view, scientists must be shown a more coherent and comprehensive explanation of history, the fossil record, and other data covered by evolutionary theory than that theory already provides. The IDM's program for bringing about change is significantly set back because its proponents have yet to explain the origin and history of the universe, Earth, life, and humanity.

Specifically, the lack of a testable model hinders ID's strategy for gaining access to public education. Efforts to persuade the Kansas School Board, Ohio Board of Education, and Dover (Pennsylvania) School Board to allow the teaching of ID concepts and to limit promotion of naturalistic biological evolution all failed. In Kansas, attempts to make instruction about radiometric dating and astronomical objects beyond the solar system "optional" (in defense of young-earth creationism) not only failed but also generated a serious backlash. In Ohio, Patricia Princehouse of Ohio Citizens for Science exulted, "We won big time here. The creationists [and ID] have lost. There is more evolution in the [educational] standards now than there would have been had they kept their mouths shut."[31] In Dover, Pennsylvania, U.S. District Judge John E. Jones III ruled that "ID is a religious view, a mere relabeling of creationism, and not a scientific theory."[32]

A TEMPTING ESCAPE ROUTE

Because the evolution/creation conflict generates such animosity, many scientists and theologians have contrived ways to circumvent the problems. The three most prominent detours are theistic evolution, God/Nature-of-the-gaps, and separate magisteria.

Theistic Evolution

Weary of the hostility and endless impasse, some scholars on both sides of the battle have suggested this simple remedy: in the classroom concede that evolution explains life's history on Earth but acknowledge the possibility of *theistic* evolution (see "Who Are the Theistic Evolutionists?" page 34). This view that God brought about the origin and development of life by working exclusively through (nonmiraculous) natural processes would seem to accommodate both religion and evolutionary science.[33]

Though it appears simple and is now taught at many Christian colleges and universities, theistic evolution introduces some ideas disconcerting to both sides of the creation/evolution debate. The notion that God had no apparent role other than to set up the natural processes offends most believers in creation. Likewise, any idea that arbitrarily interjects God into a system with no apparent need for Him offends most proponents of evolution.

Both evolutionists and creationists are quick to point out that theistic evolution offers no discernible advantage over nontheistic (naturalistic) evolution. Its explanations of life's history and predictions of future discoveries are nearly identical to those of naturalism. Thus, theistic evolution appears largely superfluous.

God/Nature-of-the-Gaps

Both theistic and nontheistic evolutionists charge that whenever believers in God encounter gaps in the knowledge or understanding of natural phenomena, they commit the "God-of-the-gaps" fallacy—claiming that's where God supernaturally intervened. Because historically several of these gaps were later discovered to have natural explanations, many theistic scientists have become reluctant to identify any perceived gap with the work of a supernatural Agent. Some insist that whether or not God exists, the record of nature is ultimately free of such gaps. Hence, they argue that while appeals to the supernatural are permissible during religious services, such appeals must remain off-limits for scientific research and education.

Ironically, a reasonable case could be made that scientists sometimes engage in similarly flawed logic. They commit what could be called the

---

Who Are the Theistic Evolutionists?

During the latter part of the 19th century and early part of the 20th, the term "theistic evolutionist" typically referred to anyone who believed that God's creation work took place over a long period of time—millions or billions of years rather than thousands. Most theistic evolutionists of that era held that God's creative involvement was not limited merely to working through natural processes and laws but included countless miraculous interventions, particularly when new species appeared on Earth.

By the end of the 20th century, however, the designation had evolved. While a few who identify themselves as theistic evolutionists still propose that God repeatedly and miraculously intervened throughout creation history, the vast majority now take a different view. Some assert that God transcended the natural order only on rare occasions—for example, at the origin of the universe, the origin of the first physical life-form, the Cambrian explosion (the sudden, simultaneous appearance of a broad diversity of complex life-forms 543 million years ago—see chapter 7, pages 138–141), and the origin of humanity. Most say that God created the universe in a "fully gifted" form. By this expression they mean that God endowed the universe at its beginning (about 13.73 billion years ago) in such a way that it would not require His miraculous intervention thereafter in order to fulfill its intended purposes (a deistic view). For the remainder of this book, unless otherwise qualified, the terms "theistic evolution" and "theistic evolutionist" refer to this 21st-century position and its proponents.

"Nature-of-the-gaps" fallacy. They presume that some unknown force or phenomenon of nature must fill *all* the gaps in human knowledge and understanding. Many theoreticians have appealed to the existence of unknown laws, principles, constants, dimensions, or hypothetical variations in the constants and laws of physics to explain a gap and dissolve supposed evidence for supernatural intervention. Stuart Kauffman and other scientists at the Santa Fe Institute, for example, appeal without any observational support or physical evidence to a hypothesized "fourth law of thermodynamics." They say that this hypothetical law would spontaneously produce a high degree of order, complexity, and information content where none exists. By this imaginative means, they explain the "natural," sudden, and very early appearance of life on Earth without benefit of a primordial soup or the arrival of prebiotics (building blocks of life molecules —see chapter 6, pages 115–123) from outer space.[34]

Similarly, some nontheistic cosmologists have appealed (without any observational support or physical evidence) to a hypothesized new law or constant of physics as a way to avoid the conclusion that the universe arose from a singularity or that the universe manifests extreme fine-tuning to allow for the existence of life.[35]

A more productive approach would be to test what happens to gaps as scientists gain more knowledge and understanding. If a certain gap becomes narrower and less problematic from a naturalistic perspective as data accumulate, then a natural explanation for the gap appears in order. However, if the gap becomes wider and more problematic from a naturalistic stance as scientists learn more, then a supernatural explanation appears in order.

The entire ensemble of gaps can be exploited to evaluate and contrast competing creation/evolution models. Such testing calls for answers to these questions:

1. Which interpretation or model contains the fewest gaps?
2. Which model(s) most accurately predict where as-yet-undiscovered gaps will be observed?
3. Which model(s) most accurately forecast what scientists will discover as they use new data and technology to explore the gaps?
4. Which model is the least contrived and most straightforward in explaining both what is known and what is not yet known?

To put this challenge another way, researchers need to make the case for their models based on factual evidence—the known instead of the unknown. The measure of a model's success must be how well that model predicts what will be discovered as researchers continue to explore current gaps of knowledge and understanding. For example, astronomers, biochemists, paleontologists, and anthropologists can take advantage of both the current body of data and current gaps in knowledge and understanding in their efforts to determine whether natural or beyond-natural explanations (and, specifically, what kind of natural or beyond-natural explanations) best account for certain phenomena relevant to the origin and history of the universe, life, and human beings.

Separate Magisteria

In one of his most famous essays, the late paleontologist Stephen Jay Gould attempted to leapfrog the entire debate. He claimed that no resolution of the creation/evolution issues is even necessary because science and religion are mutually exclusive "magisteria" (realms of knowledge and authority).[36] Science reigns supreme and alone in the realm of nature and facts, while theology reigns supreme and alone in the realm of spirit and faith. According to Gould, these two domains *do not* and *need not* overlap. Keeping them isolated means disputes or conflicts over creation/evolution problems would never arise—*if only* people wouldn't overstep the boundaries of each magisterium.

Gould's appeal resonates throughout the scientific community and even with some theologians. In his book *From Genesis to Genetics*, John A. Moore, evolutionary biologist and cofounder of the Biological Sciences Curriculum Study, described science as belonging to the domain of the "rational, demanding data and logic," while religion and theology belong to the realm of the "romantic, involving emotion, faith, and personal preference."[37] Moore labeled the evolution/creation disputes as "political disagreements, not scientific ones,"[38] because "only one side deals with science." He adds, "The theories of science have nothing to say about gods."[39] America's National Academy of Sciences (NAS) concurs.[40] In an article for science educators, NAS editors wrote, "Usually faith refers to beliefs that are accepted without empirical evidence."[41] "Science can say nothing about the supernatural."[42]

In their attempt to erect an impenetrable wall that utterly banishes references to the supernatural from discussions of cosmic and life history, some scientists opt for a redefinition of *science* as that realm of inquiry that engages in repeatable, controlled experiments producing results with determined values (and numeric error bars). Ironically, this narrow definition excludes much more than theological content. It rules out such observational sciences as astronomy and paleontology as well as such disciplines as mathematics and theoretical physics.

Science cannot hope to explain nature's record without considering observations and inferences of what happened in the past. Likewise, to ignore the relevant theoretical disciplines is to seriously impede scientific advance. While scientists, philosophers, and theologians may not always and fully agree on a definition of science, most concur that science proceeds most productively when observations and theories are used along with experiments to test competing explanations of nature's record.

The implications by Gould, Moore, and the NAS that science is all about facts, reason, and logic while religion and theology are merely romantic delusions strike some scholars as too extreme. Consequently, they have proposed various forms of slightly less separate magisteria (see "Almost-Separate Magisteria," below).

This division of truth into separate domains, objective (fact-based) and subjective (feeling-based), with science on the objective side and religious notions on the subjective side, may seem an easy way out of the conflicts. However, philosophical reality makes such a separation untenable. The very concept of truth comes from the spiritual domain. Christianity upholds the values and ethics crucial to the scientific endeavor, including support for the laws of logic.[43]

Almost-Separate Magisteria

Some scientists and theologians propose a division of science into two domains: origins science, where theological considerations are permissible, and operations (or ongoing process) science, where they are not.[44] As with Gould's "solution," this strategy creates artificial boundaries between science and theology that could limit the extent and effectiveness of both education and research. Scientific and theological testing of various creation/evolution models would certainly be hindered.

Christian theology in particular allows for a complete (or nearly complete) overlap between science and theology. From a biblical perspective, the possibility exists for varying degrees and types of supernatural intervention. Thus, both natural and supernatural causes in varying combinations need to be considered and tested across the entire record of nature. From a scientific perspective, biblical material has earned the right to consideration based on its record of accuracy in forecasting scientific discovery. The Bible states that God fills the entirety of the universe[45] and that the whole universe declares His glory.[46] The Bible also declares that God has established fixed laws to govern the heavens and earth.[47] Thus, the Bible allows for a variety of natural and supernatural causes in both origins and operations science.

These laws, particularly the law of noncontradiction (which says a proposition P is false if it asserts that both proposition Q and its denial, non-Q, are true at the same time and in the same context) allow humans to make sense of physical reality. The scientific enterprise is inconceivable without logic and mathematics. While these necessary laws of thought harmonize with a biblical worldview, they appear foreign and difficult to justify in a world wherein nature is the sole reality. Naturalism, which acknowledges no transcendent absolutes, has no necessary ground or foundation for logic, reason, or even mathematics.

COMPLETE COMPATIBILITY

In Christian doctrine, the magisteria of science and religion find total harmony. The biblical concept of faith includes confidence, based on tested/ testable evidence, in the reality of that faith's object. That is, biblical faith is fact-based. Every major Christian doctrine is either founded upon or linked to the Bible's creation accounts and hundreds of additional statements about (and descriptions of) events and purposes. Scripture gives thousands of wide-ranging scientific details on the origin, structure, and history of the universe, Earth, life, and humanity (see chapter 4, pages 66–84).

On the science side, it is difficult to find any significant description of the natural realm that does not reveal at least some element of design. Even something as mundane as electron orbits about atomic nuclei or the nuclear ground state energy levels of various elements fall within remarkably narrow limits for life's possible existence. Such apparent fine-tuning suggests at least the possibility of planning and purpose.

On the biblical side, the psalmist, for example, declares that God has revealed Himself to humanity in two books, the written record and the record of nature.[48] Both books are said to be completely reliable, having as their source the One who embodies Truth, the One who does not lie.[49] Since both, according to the Bible, reveal God and His attributes and both testify of the origins and history of the natural realm, the two books are considered not only compatible but also overlapping (see chapter 4, pages 66–68).

Perhaps an even greater problem for the separate magisteria view arises from the fact that the scientific method finds its origin in Scripture and in the work of Reformation Bible scholars (see appendix C, "Biblical Origins of the Scientific Method," pages 219–220). Despite taking courses in the history of science, few scientists seem to realize how the Bible and science work together to define this historic methodology for observing, testing, and interpreting nature's phenomena, as well as so-called revelations from nature's God. On the other hand, most Christians don't realize that the benefits of the scientific age have come from application of the standards for interpreting nature set forth in Scripture. In his letter to the Thessalonians, for example, Paul exhorted readers to put all teachings to the test and to hold on to the "good."[50] Such a practice benefits scientists and Christians alike.

A CEASE-FIRE BEGINS

Virtually every theory known to science has been established through use of the scientific method. Theories are refined, revised, or rejected by its application. As all available evidence is considered, explanations draw closer and closer to reality. Using this biblical approach to create a variety of scientific models can play a crucial role in drawing the evolution/ creation conflicts along a pathway toward resolution. If a concept belonging either to evolution or creation shows no scientific credibility and/or integrity, then that idea can and should be eliminated. Testing various scientific models will reveal more and more of the truth that corresponds to the universe and the life it contains.

CHAPTER 3

APPLYING THE
SCIENTIFIC METHOD

D iscernment kept many *War of the Worlds* listeners in touch with reality and free from panic as they heard convincing "on-the-scene" reports from the site of the supposed Martian invasion. Some turned to other radio stations to see if the same "news" was being covered. Others checked their watches and questioned the unfolding time frame. Many considered the theatrical nature of the weekly broadcast heard on that station in that time slot. Those who questioned and "tested" what they were hearing averted embarrassing and potentially dangerous reactions.

In the creation and evolution conflicts, thoughtful people need to ask questions and apply tests. Does any Darwinian model offer an adequate explanation of life's beginning and history? Does any evidence support the claim of a "young" (as in 10,000-year-old) universe and/or Earth? Do human beings possess unique spiritual characteristics or are they just advanced animals? Most important of all—is any position consistent with accumulating discoveries and increasing knowledge?

Because new information continually flows from the frontiers of research, scientific models must be refined again and again through repeated testing and the integration of relevant new data. Any model's viability depends on its ability to account for emerging discoveries. It also rests on the model's ability to specifically and accurately predict future findings.

WHAT IS THE SCIENTIFIC METHOD?

The scientific method, as drawn from the pages of Scripture (see appendix C, "Biblical Origins of the Scientific Method," pages 219–220), addresses the human tendency to form conclusions too quickly and too firmly. This systematic process guards against making any interpretation or hypothesis until certain basic steps are completed. Even after these steps are taken and a hypothesis is formed, it encourages investigators to hold loosely to their initial interpretation.

Consistent application of this step-by-step method encourages the necessary meticulousness, restraint, and humility a truth-quest warrants. Use of this process rests on and even builds on confidence that the natural realm is a well-ordered, consistent, contradiction-free system. This method and this underlying conviction, more than anything else, launched and propelled the scientific revolution of the past four centuries.

The number and wording of steps outlining the scientific method may vary slightly depending on the type of investigation to which it is applied. But its basic components—as used for explaining the origins and history of the universe, life, and humanity—include these tasks in sequential order:

1. Identify the phenomenon to be investigated and explained.
2. Identify the frame(s) of reference or point(s) of view to be used in studying and describing the phenomenon.
3. Determine the initial conditions for the phenomenon.
4. Perform an experiment or observe the phenomenon, noting what takes place when, where, and in what order.
5. Note the final conditions for the phenomenon.
6. Form a tentative explanation, or hypothesis, for how and why things transpired as they did.
7. Test the hypothesis with further experiments or observations.
8. Revise the hypothesis accordingly.
9. Determine how well the explanation of the phenomenon integrates with explanations of related phenomena.

Note: These steps apply just as usefully to biblical interpretation as they do to interpretation of natural events.

While this method does not guarantee objectivity and accuracy, it does minimize the effects of oversight, personal bias, and presuppositions. Even the most careful interpreter possesses only limited knowledge, understanding, and objectivity. Interpretations, no matter how well developed and thoroughly tested, fall short of perfection. Thus, the need for ongoing adjustments and refinements never ends. (Astronomers have measured and observed most of the solar system, but new details—even new planets the size of Pluto—continue to emerge.[1])

Therefore, this investigative method works best when practiced continuously and cyclically. Its moves researchers closer to truth, or to a more complete grasp of it, each time they cycle through the steps with new information and insight. Changes will be necessary. New questions and challenges always arise. But that's how knowledge and understanding advance. The process never ends. It's exhilarating for anyone who loves truth.

One of the strengths of this method lies in its built-in quality controls. Appropriate application eliminates bad science. Faulty or grossly incomplete interpretations will be exposed by the need for major revisions. Good interpretations yield progressively smaller revisions with each cycle through the steps. The explanatory power and predictive success of a good hypothesis continue to increase until it becomes a theory. With greater substantiation, the theory expands into a detailed and comprehensive model. Ongoing application of the scientific method continually refines, improves, and extends that model.

The scientific method makes possible a constructive, equitable dialogue about creation and evolution. All who wish to participate may present their models—interpretive scenarios developed and refined by observations and testing—for comparison. The models that demonstrate the greatest explanatory power and predictive success remain on the table for further study. Those requiring larger and larger revisions after failing predictive tests can be eliminated. The field narrows, through time, focusing on the model (or set of models) that most closely corresponds with reality. And, as the more successful models mature, they multiply into more detailed variants.

No doubt this process often involves emotion. Researchers, like other people, can become attached to their ideas, but when they bring their attachments and personal biases into the open, rather than concealing them, science can advance more freely and productively.

Within the realm of scientific research, a model's failure carries no great stigma. So-called failed models often illuminate and foster the growth of successful models. Determining what does *not* work often helps elucidate what *does*.

When all participants remain committed to follow the findings wherever they lead, they can work in partnership through failures, successes, and disagreements. They can resolve disputes and solve mysteries. Recent progress in cosmology toward a more precise and comprehensive model for cosmic origins and developments provides an illuminating example. This progress shows that truth can and does ultimately prevail on creation/

evolution questions despite ideological, philosophical, and theological preferences and commitments.[2]

A Cosmic Success Story

As the 20th century dawned, scientists entertained only one cosmological model. It was built on the belief that Newton's laws of motion explained all the dynamics of the universe and held the universe roughly static throughout infinite space and time. The infinitely old and infinitely large universe model reigned in the 19th and early 20th centuries for good reasons. This Newtonian cosmological model:

- Led to the discovery of Uranus and Neptune by successfully predicting the orbits and locations of both planets from an analysis of the orbits and locations of the six planets visible to the naked eye.
- Explained the motions and positions of all stars then visible to astronomers.
- Fit the emerging theory of Darwinian evolution by allowing for the necessary time, quantity of material, and static conditions.
- Had no competition. No other cosmic model came close to explaining so much of what scientists observed.

Figure 3.1:
Urbain-Jean-Joseph Le Verrier

In 1846, French astronomer Le Verrier noted that Uranus was slightly "out of position," according to Newton's laws of motion. After measuring the discrepancy, he hypothesized its probable cause: the existence of a planet beyond Uranus. He then calculated the position of this likely planet, based on familiar Newtonian mechanics. Just *days* later, Johann Galle at the Berlin Observatory discovered Neptune—almost exactly where Le Verrier's calculations had predicted, and scientists hailed the success of the Newtonian model of the universe. However, neither Le Verrier nor any other astronomer could explain why the point in Mercury's orbit that comes closest to the sun (the perihelion) regularly advances. The quest to solve that mystery led to development of the general theory of relativity.

(Photo courtesy of SPL/Photo Researchers, Inc.)

However, unexpected measurements made in the late 19th century cast two seemingly small shadows, or irregularities, on the Newtonian model's glowing horizon. In 1845, French astronomer Urbain-Jean-Joseph Le Verrier published his calculations showing a tiny but regular advance in the point at which Mercury's orbit comes closest to the sun (Mercury's perihelion). This advance remained after astronomers took into account all the gravitational influences of the other planets on Mercury's orbit. Then, starting in 1887, physicists began to notice that the velocity of light was not dependent upon the velocity of either the light's emitter or the observer's frame of reference. Contrary to what the Newtonian model predicted, the speed of light was constant in this context.

Albert Einstein's special theory of relativity, published in 1905, solved the puzzle of the observed constancy of light's velocity. In 1916, his general theory of relativity explained the mystery of Mercury's orbit.[3] However, the solution for these two anomalies produced a radically different cosmic model—one that specified a continuously expanding universe. This new model showed that the universe had a beginning, that the cosmos was finite with respect to time.

The expanding universe model attracted scientific attention not so much because it explained the two small anomalies, but primarily because the model predicted phenomena that astronomers had not yet seen. Given appropriate instruments and techniques that scientists either already possessed or could easily and cheaply develop, astronomers could put

Figure 3.2:
Albert Einstein

In 1916, a German-Swiss-American physicist came up with a theory to explain why the point in Mercury's orbit that comes closest to the sun keeps advancing—the general theory of relativity. Einstein's equations rocked the scientific community. They indicated that the universe had a beginning and has been continuously expanding ever since. These equations made specific predictions (about the bending of light, for example), the accuracy of which could be used to test the theory's validity. As it turns out, the predictions were confirmed. Before his death, Einstein acknowledged, based on the soundness of his theory, the necessity of a "superior reasoning power."[4]

(Photo courtesy of SPL/Photo Researchers, Inc.)

these new ideas to the test. For example, Einstein's first paper on general relativity predicted that gravity would bend space by specified amounts, which

Figure 3.3:
Sir Arthur Eddington
Mathematical physicist Eddington launched Albert Einstein to worldwide fame when in 1919, during a solar eclipse, his British research team observed and measured the bending of starlight (by the sun's gravity). The degree of bending matched what Einstein's equations had predicted. Eddington joked that only two people in the world understood Einstein's theory and he was not too sure about the other fellow. For much of his life he tried (unsuccessfully) to find some formulation of relativity that would get around the need for a cosmic beginning.

(Photo courtesy of SPL/Photo Researchers, Inc.)

observations could either verify or refute.[5]

The success of Einstein's predictions, along with other measurements and observations made possible by technological advance, so convincingly affirmed the universe's continuous expansion that astronomers began, if somewhat reluctantly, to abandon the Newtonian model. Britain's famous mathematical physicist Sir Arthur Stanley Eddington gave voice to that reluctance. He described the concept of a cosmic beginning as "philosophically repugnant"[6] because it would not "allow evolution an infinite time to get started."[7]

Eddington reacted to the theistic implications inherent in a beginning, or creation, of the universe. In addition, he, like most scientists, was motivated to salvage as much of the old science as possible. (Consider how construction workers react when a city inspector arrives during the building of a new structure only to say, "Stop the work. The foundation and framing do not meet city codes." Their first thought, after a big groan, is to repair what's faulty without tearing down the entire structure.) Consequently, Eddington and many other astrophysicists attempted to adjust Einstein's model. They desired a modification that would still "allow evolution an infinite time to get started."[8] They proposed three classes of cosmic models:

1. The hesitation model—a universe that expands, but through a hypothesized new constant of physics and a careful choice of the constant's sign and value slows its expansion rate to zero, and a nearly infinite time later starts to expand again.
2. The steady state model—a universe that continuously expands but with a hypothesized "creation force" always bringing new matter into existence so that the universe maintains the same conditions for infinite time.

3. The oscillating model—a universe that alternates, thanks to hypothesized new physical mechanisms and hypothesized new constants of physics, between expansion and contraction for infinite time.

Each of these three sets of cosmic models was founded on different assumptions, hypotheses without any supporting physical evidence. Normally, the ad hoc nature of such models would have resulted in immediate rejection by the rest of the scientific community. However, many nontheistic scientists considered the avoidance of a cosmic beginning a strong reason to give these models careful consideration.

More importantly, these models were retained for investigation because each came with its own set of predictions about what astronomers might find as they looked more deeply and precisely into the cosmic past. When put to test after test, however, none of the observations matched the predictions. This result eventually led astronomers to discard their proposed adjustments and to focus on a narrower cluster of the three sets of cosmic models. The models in this cluster incorporated Einstein's unwanted (single) cosmic beginning in finite time:

1. Cold big bang models (proposing that the universe expands from an infinitesimally small but cold volume)
2. Hot big bang models (proposing that the universe expands from an infinitesimally small and nearly infinitely hot volume) dominated by ordinary matter—protons, neutrons, and electrons—that strongly interacts with radiation
3. Hot big bang models dominated by exotic matter—particles such as neutrinos, axions, and neutralinos—that weakly interacts with radiation

These models *did* fit the observations, at least initially.

As researchers continued to make observations and conduct tests, they eventually ruled out the first two sets of models. But new tests and observations reinforced the plausibility of the third set of models. Astrophysicists by now have established an array of amazingly detailed models that fit within the third category.

Today, the viable models are hot big bang models in which "dark energy" (the self-stretching property of the cosmic space fabric) dominates "exotic dark matter" (particles that do not strongly interact with photons), which in turn dominates "ordinary dark matter"[9] (aggregates of protons, neutrons, and electrons that do not emit appreciable light)—see table 3.1, below. The proliferation of new models stays within an increasingly narrow range. In other words, researchers continue to make remarkable progress toward resolving their scientific differences—despite philosophical or religious preferences. For now, they vigorously and creatively pursue the truth about cosmic origins wherever it leads.

TABLE 3.1: COSMIC INVENTORY

Princeton cosmologists Masataka Fukugita and Philip James Edwin Peebles, after an exhaustive compilation of the best measurements made on the density components of the universe, assembled the following inventory of the universe:[10]

COSMIC COMPONENT	PERCENTAGE OF TOTAL COSMIC DENSITY*
Dark energy (self-stretching property of cosmic space surface)	72%
Exotic dark matter (particles that weakly interact with light)	23%
Ordinary matter (protons and neutrons)	4.5%
Stars and stellar remnants	0.27%
Planets	0.0001%

* The missing 0.2299% is part of the uncertainties for dark energy and exotic dark matter. Both are known to only 2% accuracy.

RELIGIOUS FREEDOM IN RESEARCH

Cosmology research has been enhanced by the openness of prominent participants to state their philosophical and theological perspectives. In proposing new models (or assessing existing ones), Sir Fred Hoyle, John Gribbin, Robert Dicke, Philip James Edwin Peebles, Fang Li Zhi, Li Shu Xian, Arno Penzias, and Stephen Hawking, among many others, have openly declared their theological motivations and leanings (see "Theological Comments from Cosmologists," page 50–51).

Figure 3.4: The Hubble Ultra Deep Field
This image from the Hubble Space Telescope shows the deepest view of the visible universe yet achieved. The width of the field is about a tenth of the (angular) diameter of the moon, and the image required a 278-hour exposure. It reveals the first galaxies ever to form (when the universe was only about 400 million years old). From the number of galaxies seen in this image, astronomers estimate that the universe contains at least 200 billion galaxies. The Hubble Ultra Deep Field shows that galaxies far away are much closer together than galaxies up close. This spreading apart of the galaxies through time provides direct evidence of the continuous expansion of the universe from an infinitesimal volume.

Theological Comments from Cosmologists

Sir Arthur Stanley Eddington

Philosophically, the notion of a beginning of the present order of Nature is repugnant to me....I should like to find a genuine loophole.[11]

We [must] allow evolution an infinite time to get started.[12]

Sir James Jeans

It is difficult, but not impossible, to believe that matter can be continuously in process of creation.[13]

Sir Fred Hoyle

This [steady state] possibility seemed attractive, especially when taken in conjunction with aesthetic objections to the creation of the universe in the remote past. For it seems against the spirit of scientific enquiry to regard observable effects as arising from "causes unknown to science."[14]

Sir Hermann Bondi

The problem of the origin of the universe, that is, the problem of creation, is brought within the scope of physical inquiry, and is examined in detail instead of, as in other theories, being handed over to metaphysics.[15]

John Gribbin

The biggest problem with the Big Bang theory of the origin of the Universe is philosophical — perhaps even theological — what was there before the bang? This problem alone was sufficient to give a great initial impetus to the Steady State theory, but with that theory now sadly in conflict with the observations, the best way round this initial difficulty is provided by a model in which the universe expands from a singularity, collapses back again, and repeats the cycle indefinitely.[16]

Robert Dicke

The matter we see about us now may represent the same baryon [i.e., matter] content of the previous expansion of a closed universe, oscillating for all time. This relieves us of the necessity of understanding the origin of matter at any finite time in the past.[17]

(continued)

Philip James Edwin Peebles

This conventional theory [the big bang] has two defects: it fails to explain why the large-scale mass distribution should be so close to uniform and it requires that the expansion can be extrapolated back to a singularity at a definite time in the past.[18]

Arno Penzias

Astronomy leads us to a unique event, a universe which was created out of nothing, one with the very delicate balance needed to provide exactly the conditions required to permit life, and one which has an underlying (one might say "supernatural") plan.[19]

Stephen Hawking

It would be very difficult to explain why the universe should have begun in just this way, except as the act of a God who intended to create beings like us.[20]

By and large, cosmologists realize that their research provides a crucible for testing various philosophical and theological constructs. These concepts may or may not prove consistent with established scientific findings. No religion or philosophy remains fully insulated from (or, for that matter, personally insulted by) observational tests. In the words of China's most famous astrophysicist, Fang Li Zhi, "A question that has always been considered a topic of metaphysics or theology, the creation of the universe, has now become an idea of active research in physics."[21]

In practicing religious freedom, cosmologists grant to one another the right to use observations and measurements to demonstrate which theological propositions do or do not correspond with the observable universe. In principle, their desire for truth outweighs the desire to cling to an ideology.

A TACTICAL CHANGE

The shift in cosmology theory over the past century demonstrates how science replaces a vague or inaccurate concept of nature (in this case, the origin and subsequent development of the universe) with a more precise model. Typically, researchers hold fast to an existing model, no matter what its defects, until they see an alternate model that works better in five ways.

The new model must (1) give a wider and more detailed view of what's going on; (2) make better sense of the established data; (3) provide more reasonable and consistent explanations for the phenomena under investigation; (4) result in fewer unexplained anomalies and gaps; and (5) prove more successful in anticipating or predicting future findings.

Science acknowledges that every model is to some extent inadequate and imperfect. However, knowledge can and does advance when a new approach explains more, explains better, and predicts more successfully than does its predecessors. Eventually these advances get passed on to students—first at the graduate level, then undergraduate, then secondary, and finally elementary. By the time it reaches the elementary level, the new framework has usually filtered into general awareness.

Though still somewhat slow moving, this trickle-down effect proceeds more rapidly today than it has in past decades because of advances in communication. But that same technology also spreads misinformation and disinformation, giving educators good reason to be even more cautious than researchers. Teachers and curriculum providers are not likely to "go boldly where no one has gone before" until they see considerable public consensus among research scholars and until new models become well established.

Building RTB's Model

Scientists may cling to evolutionism and naturalism because, as yet, they have not been presented with better scientific models. Likewise, some Christians may cling to young-earth creationism or theistic evolution because, as yet, they have not seen better biblical creation models. Use of the biblically based scientific method and presentation of testable models brings hope for tempering hostilities on the creation/evolution battle-fronts. Developing models that offer better and more thorough explanations could possibly bring warring parties to the same side of the table. It might also bring frustrated bystanders into more active and supportive roles in either the scientific or the Christian enterprise—or both.

This book outlines a model that strives to uphold both scientific and biblical integrity as it attempts to reconcile the goals of the scientific community with the goals of the Christian community. This model is testable, falsifiable, and predictive. In presenting its model, RTB and its scholar team hope not only that creation/evolution conflicts will de-escalate but also that all sides will recognize that significant progress toward resolution can be achieved.

As with all models, RTB's creation model can only benefit from competition—the more models to come forward (either for evolution or creation), the better. The best of them will undergo revision and refinement as test results accumulate and research advances. As long as participants remain committed to following the truth wherever it leads, they can work in partnership through failures, successes, and disagreements to resolve disputes and solve the mysteries that surround the origins and histories of the universe, Earth, life, and humanity.

Biblical Material

This book presents the RTB creation model as a scientifically viable response to creation/evolution battles and impasses. It arises from a theistic hypothesis (see step 6 in the scientific method, page 42) that comes from one specific source, the Bible. It was not derived from any other religious text or from vague intuition. Every element of this biblically based hypothesis, as it grew step-by-step into a model, was drawn from, and attempts to remain consistent with, the data in all 66 books of Scripture.

RTB's specific biblical perspective, while acknowledging that a considerable portion of the Bible's creation material contains metaphors and figurative language, nevertheless holds that the Bible's creation content is predominantly literal in its descriptions of natural phenomena. Its accounts are viewed as reliably factual in their declarations about the origin, history, chronology, and current state of both the physical universe and life within the universe. In these respects, RTB's model remains consistent with the creation tenets of the Reformation creeds.[22]

A biblical model's validity depends partly upon the soundness of human attempts to interpret the words of the Bible. This model also depends on the soundness of the efforts made to integrate those theological understandings with human interpretations of the natural data (science). Any weaknesses and errors are assumed to come from unsound interpretations of one or both sets of data, *not from the source material itself.*

Accuracy also depends on the degree of completeness. The RTB creation model attempts to take into account every biblical passage with known relevance to creation and evolution. (A list of all such references appears on the RTB Web site.[23]) The major biblical creation accounts and their themes are listed in table 3.2, page 54.

Chapter 4 discusses the biblical foundation for RTB's creation model in some detail.

TABLE 3.2:
THE MAJOR BIBLICAL CREATION ACCOUNTS

REFERENCE	THEME
Genesis 1	Creation chronology: physical perspective
Genesis 2	Creation chronology: spiritual perspective
Genesis 3–5	Human sin and its damage
Genesis 6–9	God's damage control
Genesis 10–11	Global dispersion of humanity
Job 9	Creator's transcendent creation power
Job 34–42	Creation's intricacy and complexity
Psalm 8	Creation's appeal to humility
Psalm 19	Creation's "speech"
Psalm 65	Creator's authority and optimal provision
Psalm 104	Elaboration of physical creation events
Psalm 139	Creation of individual humans
Psalms 147–148	Testimony of the Creator's power, wisdom, and care in nature
Proverbs 8	Creator's existence before creation
Ecclesiastes 1–3; 8–12	Constancy of physical laws
Isaiah 40–51	Origin and development of the universe
Romans 1–8	Purposes of the creation
1 Corinthians 15	Life after life
2 Corinthians 4	Creator's glory in and beyond creation
Hebrews 1	Cosmic creation's temporality; role of angels in creation
2 Peter 3	Creation's end
Revelation 20–22	The new creation

Nature's Material

Since Darwin attempted to explain life's history, the body of relevant research findings has multiplied many times. In some scientific disciplines, the knowledge base doubles every five years or less. These new facts fuel optimism for the rapid emergence of better models.

However, the sheer enormity of the information explosion poses a problem. Researchers are forced to focus on increasingly specialized subjects. The task of integrating all the scientific data, even of selecting what seems most important for understanding the creation/evolution story, appears daunting. An enormous team effort is required for RTB's model-building process. The results are contained not just in one volume, but rather in several, with more to come.[24]

No one book could possibly set forth all the facts and information necessary to build a scientific model depicting the biblical account of the cosmos's and life's origin and history. This chapter offers an abbreviated list—86 of the most essential realms of data from the record of nature that must be accounted for by a reasonably complete creation/evolution model. Most of the data come from the physical and life sciences. A few come from the humanities. Any viable model must either demonstrate how it anticipates and explains each item in the data list or show valid, testable reasons why an item is irrelevant or false. The references indicate where more extensive explanations of RTB's creation model relevant to each particular item or topic can be found.

SCIENTIFIC DATA FOR CREATION/EVOLUTION MODELS TO EXPLAIN

1. The beginning of the universe about 13.73 billion years ago[25]
2. The beginning of the cosmic space-time configuration[26]
3. The transcendence of the cause for the beginning of space, time, matter, and energy
4. Adequate dimensionality, in addition to length, width, height, and time, to account for the pervasive coexistence of gravity and quantum mechanics (a total of 10 dimensions in all)
5. The distribution of matter and energy across the cosmic space-time surface
6. The precise values of the physical laws and constants necessary to make physical life possible[27]
7. The constancy of the physical laws and constants[28]
8. The exquisitely fine-tuned continuous cosmic expansion essential to life[29]
9. Continuous cooling of cosmic background radiation from near infinite temperature
10. Relative abundances of the elements before the advent of stars
11. Buildup and relative abundances of nonradiometric elements heavier than helium

12. History of the relative abundances of radiometric isotopes
13. The size of the universe[30]
14. The cosmic dominance of darkness rather than light
15. The location and abundance of exotic matter relative to ordinary matter
16. The anthropic principle[31]
17. The anthropic principle inequality[32]
18. Humanity's unique (and ideal) location for viewing the cosmos[33]
19. Humanity's unique (and ideal) time window for viewing the cosmos[34]
20. Cosmic timing of the peak abundances for uranium and thorium
21. Timing and locations of supernovae occurrences at Earth's solar system origin site[35]
22. The timing and all other details of an early collision between a Mars-sized body and Earth[36]
23. The effects and timing of the Late Heavy Bombardment[37]
24. The lack of evidence for possible ETI
25. The uniqueness of Earth's solar system among all observed planetary systems
26. The absence of life on other solar system bodies
27. The unique life-essential properties of the Earth-Moon system
28. The decline (by a factor of six) in Earth's rotation rate
29. The decline (by a factor of five) in Earth's heat from radioactive decay
30. The stability and features of Earth's terrestrial magnetic field and internal dynamo
31. Earth's plate-tectonic history
32. The lack of a natural source for terrestrial or extraterrestrial prebiotics
33. The lack of a natural source for homochiral building blocks of biomolecules
34. The (early) timing of life's origin[38]
35. The suddenness of life's origin[39]
36. The complexity and diversity of Earth's first life
37. The lack of a primordial soup[40]
38. The predominance of bacterial life for the first 3 billion years of life's history
39. The ubiquity and complexity of biochemical design[41]
40. The ubiquity and complexity of biochemical organization[42]
41. The history and diversity of Earth's sulfate-reducing bacteria[43] and cryptogamic colonies[44]

42. Earth's oxygenation history[45]
43. The sun's dimming (by 15 percent) during its first 1.5 billion years[46]
44. The sun's brightening (by 15 percent) during the past 3 billion years
45. The unusual current stability of the sun's luminosity
46. The pattern of advances in life's complexity
47. The history and abundances of water-soluble elemental (but life-essential) poisons
48. The Cambrian explosion[47]
49. The frequency and magnitude of mass speciation events[48]
50. The frequency and magnitude of mass extinction events[49]
51. The rapidity of life's recovery from mass extinction events
52. The Lazarus taxa phenomenon[50]
53. The occurrence of biological and biomolecular convergence events (repeated evolutionary outcomes)[51]
54. DNA similarities among diverse species[52]
55. The sudden recent cessation of speciation[53]
56. The timing of Earth's petroleum production peak[54]
57. The timing of Earth's petroleum storage peak[55]
58. The history and frequency of multiple-species symbiosis[56]
59. The emergence of a self-preservation drive
60. The emergence of the uniquely human drive for meaning (a sense of hope, purpose, and destiny)
61. The emergence of "soulish" behavior as expressed in higher animals
62. The emergence of "spiritual" behavior as expressed in humans
63. The quantity and diversity of expressions of altruism in nature
64. The social structure and division of labor among insects
65. The timing of vascular plants' origin and proliferation
66. The ubiquity and diversity of carnivores and parasites[57]
67. The longevity of various species in the fossil record
68. The abundance of transitional forms among large-bodied, low-population species[58]
69. The scarcity of transitional forms among small-bodied, large-population species
70. The ubiquity of optimized ecologies
71. The rapid development of optimized ecologies[59]
72. Apparent "bad designs" in complex organisms[60]
73. The absence of "bad designs" in simple organisms and inorganic structures

74. The life spans of various species
75. DNA differences among humans, Neanderthals, and chimpanzees
76. The low population levels of nonhuman hominids
77. The timing and other characteristics of humanity's origin[61]
78. "Big bangs" in the arrival of jewelry, art, technology, clothing, communication, and culture
79. Changes in the human life span[62]
80. Humanity's unique characteristics and capabilities[63]
81. Humanity's over-endowment for basic survival[64]
82. The timing and locus of human origin
83. The descent of modern humans from one man or a few men and from one woman or a few women
84. The narrow physical limits on the time window for human civilization[65]
85. The broad physical limits on the time window for simple life-forms
86. The physical laws' optimization for restraining evil

Born of Science and Scripture

Several commitments infuse RTB's model with the potential for growing explanatory power and scope. RTB is committed to keep or maintain:

- Simplicity first, which implies that what is incompletely or poorly understood, inadequately tested, and highly complex will be interpreted in light of that which is simple, better understood, and more thoroughly tested. (See "Why Simple Sciences First?" page 59.)
- The big picture (explanations for the overall phenomenon) ahead of attempts to interpret and understand the minute details
- Biblical integrity—honorable application of hermeneutic principles to interpreting and integrating all biblical data
- Scientific integrity—honorable application of the scientific method to well-established, observationally tested data
- Humility—honorable respect for nature and students of nature, for Scripture and students of Scripture, and for human limitations
- Continual improvement, which implies that since there is always more to learn and understand about any phenomenon, all models need to be constantly updated, revised, and improved
- Diligence and creativity in generating predictions (based on the model) of what scientists and theologians will discover in their future research endeavors

One additional RTB commitment is the most important of all—a commitment to follow, regardless of personal cost, wherever the evidence leads.

Why Simple Sciences First?

Simple sciences are not "simple" in that they are easy to understand. In contrast to complex sciences, simple sciences are those disciplines in which the phenomena being studied can be well defined by equations of state. Differential equations effectively describe the behavior of these phenomena. Because of such simplicity, conclusions about possible causal agents are subject to much more rigor and show less ambiguity. In this regard, philosophical and theological conclusions may be more easily drawn from the simple sciences than from the complex.

Specifically, RTB's scientific case for the God of the Bible as the Designer and Creator of all nature starts with mathematics, proceeds to astronomy and physics, moves on to planetary science and geophysics, progresses to geology and chemistry, advances to biochemistry and microbiology, goes on to botany and zoology, and ends at anthropology. Building the model in incremental steps aids the integration process.

Taking small steps toward the complex sciences allows for progressive evaluation of the evidence—to test whether the indications of a supernatural, caring, intelligent Designer grow progressively stronger or progressively weaker. If compelling evidence for the God of the Bible becomes discernible in the simple sciences, then it should be discoverable in the complex sciences as well—with two important provisos: (1) The database must be sufficiently complete, and (2) testing must be thorough enough and persistent enough to reveal evidence of design hidden (initially) within the complexity.

RTB's commitments should reassure skeptics on all sides—those who claim the metaphysical doesn't exist, those who complain they won't be allowed to ask hard questions of the models, those who worry that the Bible's message may not be given due weight and consideration, and those who worry that the achievements of science and scientists will be belittled. In other words, these commitments can assuage the fears of those who worry that dogmatism of one kind or another will trump the truth. A consistent demonstration of these commitments in the model-building process can go far, at least for most inquirers, toward dispelling doubts and concerns.

Part II of this book moves toward restoration of the partnership and productivity once enjoyed by scientific and biblical researchers as they consider some of life's greatest questions. It outlines a model for the origin and history of the universe and life that is scientifically, philosophically, and theologically integrated. The model is fairly comprehensive in scope, though RTB acknowledges the need for the model's expansion and improvement.

One reason for the public's documented suspicion of the more prominently exposed creation/evolution models is that the presuppositions and research principles undergirding such models are either hidden or incompletely spelled out. The RTB creation model differs in this important respect: the biblical, theological, and philosophical presuppositions on which the model rests are specifically stated for all to see. They can be falsified or further affirmed through ongoing investigation.

While some creation/evolution models attempt to explain just the biological realm, the geophysical realm, or the cosmic realm, the RTB creation model attempts to integrate and address all these realms. And rather than focus strictly on nature's what, where, and when, this model also tackles the how and the why. Though serious in its offer to provide the best (to date) answers to questions about the origin and history of the universe and life—answers that correspond to physical and spiritual reality—the RTB creation model also serves as an example. It is presented in the hope of encouraging development of competing, equally comprehensive creation/evolution models. Competition can generate refinements at a faster pace than is possible when models stand alone.

The next six chapters present just the highlights of the RTB creation model and demonstrate how the model fares in light of emerging research findings. Readers with little interest in the scientific details of the model may prefer to skim these chapters and move ahead to chapter 10.

The model's highlights are organized as follows:

- Chapter 4 describes the most significant biblical, theological, and philosophical foundations of the RTB creation model.
- Chapters 5 through 9 test the validity of these foundations and show the model's resilience in accounting for the scientific advances of the past few decades. These advances have actually served to enhance the model's explanatory power.
- The material in chapters 5 through 8 progresses from the simpler science disciplines to the more complex—that is, from mathematics, astronomy, and physics to geology and chemistry to biology, anthropology, and sociology. They also transition from the big picture view to a smaller scale—from the universe as a whole, to the Milky Way Galaxy, to the solar system, to Earth, to Earth's surface, and to the cities, farms, mines, and resorts in which

humans live, work, and play. Not only do these chapters consider questions of causation, but they also discuss the intriguing how and why questions.

- Chapter 9 addresses what some skeptics may describe as the most difficult and complex challenges to the RTB creation model.
- The final three chapters suggest ways it can be tested further through the success or failure of its research predictions and through development of new testing standards, tools, and practices.

Part II

BUILDING

THE

RTB

CREATION MODEL

CHAPTER 4

THE BIBLICAL FRAMEWORK OF THE RTB CREATION MODEL

O ne key contributor to the panic generated by *The War of the Worlds* broadcast was the Mercury Theatre actors' skill in replicating the authoritative tone of the evening news. People expect reporters to give a truthful account of what's happening. Orson Welles' skillful simulation of a newscast took advantage of listeners' tendency to trust the authenticity of the nightly report.

The Bible claims to be a source of truth, and for centuries it was considered by many people, especially in the West, a more reliable source than any other. Then "news" reports from the sciences seemed to challenge and even undermine its credibility. The much less familiar news story, summarized in chapter 3, tells how the past century's research reversed some of that erosion by showing the accuracy of the biblical material on cosmology. This story became the subject of a number of books, including one by Harvard astronomer Robert Jastrow.[1] In his words,

> For the scientist who has lived by his faith in the power of reason, the story ends like a bad dream. He has scaled the mountains of ignorance, he is about to conquer the highest peak, as he pulls himself over the final rock, he is greeted by a band of theologians who have been sitting there for centuries.[2]

As Jastrow observed, the Bible anticipated by thousands of years the scientific discoveries of the big bang's hallmark features.

In light of this remarkable feat, to investigate the biblical data on other aspects of the natural realm seems promising. Accumulating evidence for the origin and history of the Milky Way Galaxy, the solar system, the planet, and Earth's life-forms, including humanity, may likewise be anticipated by the biblical account of creation and evolution.

Because the Bible contains such an abundance of commentary on the natural realm (about 20 major passages and hundreds of additional references), it supplies ample material from which to construct a scientifically testable model. Far from being just another ancient Near Eastern creation myth, Genesis 1 is a sweeping, plausible narrative, and its structure, a source of the classic scientific method (see chapter 3, pages 42–44, and appendix C, "Biblical Origins of the Scientific Method," pages 219–220). Each of the RTB biblical model's features emerged from this method's application.

ANCHORING THE MODEL'S FOUNDATION

The RTB creation model rests upon four cornerstone inferences drawn from Scripture:

- Dual Revelation: the Creator's attributes (truth, love, wisdom, power, and so on) ensure the harmony of His creative works (nature) with His inspired Word.
- Purposes of Creation: nature fulfills the Creator's stated and implied purposes.
- A Wide-Angle View of Creation Chronology: the Creator transcends the creation; the realm of nature has a beginning and an end, a *before* and an *after*.
- Detectability of Divine Intervention: the Creator's interventions can be ascertained via close and careful study of the creation (nature).

These four inferences lay a foundation for every piece, large and small, of the RTB creation model. Given their structural significance, the four cornerstones will be established first. These cornerstones provide the context for understanding some of the most intriguing "Characteristics of Creation" (see pages 72–80), the various beams in the model's frame. The chapter ends with a brief commentary on the Creator's options and the latitude of biblical language, then an overview, in outline form, of the more significant biblical/theological components of the RTB creation model.

DUAL REVELATION

The Bible repeatedly declares that God is truthful and does not lie in word or in deed.[3] He is unchanging, not capricious;[4] He is almighty, not just mighty.[5] He is the personification of wisdom and of love.[6] Therefore, according to Scripture, God's revelation both through the record of nature and through the words of the Bible can be considered trustworthy and free of contradiction and error.

One of the great Reformation creeds, the Belgic Confession, describes this crucial cornerstone in these words:

> We know him [God] by two means: First, by the creation, preservation, and government of the universe, since that universe is before our eyes like a beautiful book in which all creatures, great and small, are as letters to make us ponder the invisible things of God: his eternal power and his divinity, as the apostle Paul says in Romans 1:20. All these things are enough to convict men and to leave them without excuse. Second, he makes himself known to us more openly in his holy and divine Word, as much as we need in this life, for his glory and for the salvation of his own.[7]

The RTB creation model thus infers that the record of nature and the words of the Bible will always be in harmony with one another, with no possibility of contradiction. However, science is not the same as the record of nature. The disciplines of science involve human interpretation. In some instances these interpretations can be faulty and/or incomplete. Similarly, Christian theology is not the same as the words of the Bible. Like science, theology involves human interpretation, which may be faulty and/or incomplete. So conflicts between science and theology do not necessarily falsify or damage the RTB creation model. Rather, such discord can provide opportunities for exposing incomplete understanding and faulty interpretations and for discovering a deeper or more complete harmony.

PURPOSES OF CREATION

The Bible says that no finite mind can fully grasp all of God's reasons for creating the universe, the earth, and all life—and for creating them in the manner He chose.[8] The apostle Peter comments that even angels lack full understanding of God's purposes in creation.[9] Nevertheless, the Bible explicitly states several purposes for God's creating things as He did and implies others, including His intention to do the following:

- Express His attributes, specifically His glory, righteousness, majesty, power, wisdom, and love
- Provide a suitable habitat for all manner of physical life and for human beings in particular until the stage is set for a brand-new creation
- Relate to human beings, revealing to them not only His glory but also the wonders and weaknesses of their nature
- Supply resources for the rapid development of civilization and technology and the achievement of global occupation
- Provide humanity the best possible viewing conditions for discovering, through a careful examination of the cosmos, His existence and attributes
- Set up the optimal physical theater—including an optimal human life span—for conquering sin and the evil it produces
- Demonstrate in person His attributes, specifically His glory, righteousness, power, wisdom, and love
- Offer all people a way to live forever in His presence in a new creation—apart from sin and death and every consequence of rebellion against Him
- Equip all who accept that offer of immortality to fulfill their role in the present creation and to enjoy their reward in the new creation

Even this limited list of God's purposes for creating help answer some of the more troubling *why* questions confronting a biblical, specifically Christian, worldview—questions such as "Why would a good and all-powerful God create a universe that yields so much human suffering?" Though God could conceivably create a more pleasant and less onerous planetary home for humanity, at least temporarily, God has greater purposes for the universe and Earth than simply to serve as a relatively comfortable human habitat. The physical features of the universe and Earth may indeed be ideal, if one considers the bigger picture.

A WIDE-ANGLE VIEW OF CREATION CHRONOLOGY

One critical distinctive of RTB's creation model is its perspective on the cosmos as part of a larger reality. In a linear reference to time, there's a *before* and *after*. Genesis and other passages hint at what existed *before* God created this universe. Revelation, along with other texts, describe what will come *after* it—an entirely new creation.

The most significant feature of the new creation to come is that by choice, God and His people will live together face-to-face without the effects of decay, pain, suffering, evil, and darkness (physical and spiritual) yet will still be fully capable of personal, purposeful, loving action and interaction. The new creation is not merely the paradise of the Garden of Eden (Genesis 2) restored. It is a radically different creation governed by different laws and framed by different dimensions.[10] The promise of a completely new creation is a critical distinctive of the biblical worldview, a belief that separates the Bible's creation story from all other creation and/or evolution stories. It thus sets apart RTB's creation model from other models of the origin and history of the universe and all life.

The Bible states both implicitly and explicitly that the present creation serves as preparation for the new creation. God uses this current, familiar realm as a platform to reveal Himself and to draw people to Himself—through Jesus, the incarnation of God—across an impassable divide caused by willful human autonomy. According to the Bible, every-thing in and about the current creation in some way contributes toward this larger purpose, as outlined above. The creation chronology summa-rized below represents a sequential outline of the RTB creation model:

1. God, who exists outside space and time, caused the universe of matter, energy, space, and time to come into existence.
2. God guided the expansion and cooling of the universe toward the formation of a suitable planetary home for humanity.
3. God successively transformed Earth and the solar system through six major creative stages in preparation for human habitation. Throughout these creation periods God successively layered increasingly advanced plant and animal life to maximize support for humanity's global expansion and civilization.
4. God created Adam, the first human, and placed him in Eden. God taught Adam to care for this magnificent garden, then created Eve to be his mate and helper.
5. God allowed Lucifer, once the most beautiful and powerful of his angelic creatures and then the perpetrator of a rebellion against God in the supernatural realm, to enter Eden. There, God permitted Lucifer to entice Adam and Eve toward autonomy and self-exaltation.
6. Adam and Eve defied God's authority and thereby introduced sin (and the evil it produces) into the earthly environment.
7. God banished Adam and Eve from Eden, removed their access to physical immortality (the tree of life), and later shortened the life span of their descendants. These actions restrained the expression of evil and made way for redemption.

8. God chose to communicate to humans—through Abraham and his progeny—specific instruction and preparation for His plan to redeem humanity from their hopeless plight. This plan and God's promise to fulfill it gave humankind hope.
9. God came to Earth as a human (Jesus Christ), resisted temptation to sin, and paid sin's death penalty on humanity's behalf. Motivated by love, Jesus endured death on a cross so that anyone, through faith, may receive justification before God, fellowship with Him, and the capacity to grow in godliness. This provision includes the promise of eternal life with God.
10. Jesus Christ conquered death, proved His resurrection by appearing physically to hundreds of eyewitnesses, and then ascended into the transcendent realm (heaven) where He prepares the new creation for all who accept His provision.
11. Jesus Christ commissioned His followers to spread the good news about His offer—with gentleness, respect, and a clear conscience—to every person on the planet.
12. Jesus Christ will remove from His presence all who refuse His gift and reject God's authority. Jesus will escort His people into the new creation, where they will live forever face-to-face with Him. Once there, His people will fulfill new leadership roles in His service.

This chronology delineates the time frame for RTB's model. This time frame, as it relates to the fulfillment of God's purposes for creation, is the key to understanding the various features of that model.

What's So Important About a *Before* and an *After*?

The existence of a reality before the universe and beyond the universe carries enormous philosophical implications—and practical ramifications. Woody Allen seems to capture the point. In his movie *Annie Hall*, a mother takes her school-aged son to the family doctor for help with changes in his mood and behavior. When asked what is bothering him, the boy tells the doctor that the universe is expanding, which means it will one day break apart. Since coming to this realization, he has seen no reason to do his homework. "What's the point?" he asks. The doctor has no answer.

In *Stardust Memories*, Allen, taking the role of a filmmaker, glibly asks his associates, "Hey, did anyone read on the front page of the *Times* that matter is decaying?" In the next few lines he explains the significance of this news—"The universe is breaking down.... Soon there's not going to

(continued)

be anything left." The dark irony he conveys in comedic fashion is that their enterprise, in fact their very existence, leads to nothing. Life is essentially meaningless.

A research paper published in *Astrophysical Journal* conveys the same message. Astrophysicists Lawrence Krauss and Glenn Starkman point out that, according to recent discoveries, "dark energy" will cause the universe to continuously expand at an accelerating rate.[11] Eventually it will fly apart too far and too fast for metabolic reactions to occur. At that point, life everywhere in the universe will die. With the death of all life comes the end of all consciousness. With the end of all consciousness comes the end of any real hope, purpose, or destiny.

If evolutionists are correct in assuming that there is no *before* or *after* to the universe, then the schoolboy and the filmmaker are right to be depressed. They are the ones in touch with reality. However, if the universe is not "all that ever was, is, or will be,"[12] the boy can find a reason to do his homework and the producer can discover a reason to create films.

DETECTIBILITY OF DIVINE INTERVENTION

The entire chronology of God's plan for humanity rests on the *reality* of a supernatural realm. Though it may seem otherwise, belief in that reality is not merely subjective — nature provides observable, measurable verification, as in the case of cosmology. The Bible claims that creation began in a miracle and that it unfolds in a miraculous way for the benefit of humanity, sustained moment by moment in the constancy of the Creator's care. (See chapters 5–9.) The Bible describes three types of miracles:

Transcendent miracles — This type of miracle involves acts of God performed outside or beyond the limitations imposed by the laws of physics and the space-time dimensions of the universe. Study of Scripture indicates obvious examples from the natural realm such as the creation of space-time dimensions, establishment of the physical laws, and creation of humanity's spiritual nature.

Transformation miracles — These miracles involve God's direct actions. In them He refashions certain aspects of His created realm or takes components within it to manufacture something of much higher complexity and functionality. Transformational miracles take place within, not outside, the laws of physics and space-time dimensions of the universe. Here, God works with what already exists in a manner that produces results far beyond what natural processes, by themselves, could reasonably be expected to produce. Two examples are the origin of purely physical life (see chapter 6, pages 115–123) and the fashioning of Earth and its continents and oceans with the exact features necessary to sustain advanced life on an abundant, diverse, and global scale (see chapter 7, pages 128–143).

Sustaining miracles—The Bible says that God works continuously throughout cosmic history to ensure that everything in the universe maintains the just-right conditions for support of human life. According to Colossians 1:16-17, "For by him [Christ] all things were created: things in heaven and on earth, visible and invisible; . . . and in him all things hold together."

If this statement is true, scientists can expect to discover that the laws and constants of physics as well as all discoverable characteristics of the universe, of Earth, and of life manifest exquisite fine-tuning.

According to the Bible, God performs relatively few miracles of the transformational or transcendent types. He does only those miracles necessary to achieve His purposes. The Bible further indicates such miracles are episodic—short periods when several or many transformational and/ or transcendent miracles may occur separated by long time spans during which no such miracles take place. Also, the biblical pattern shows that transformational miracles far outnumber transcendent miracles.

This rarity of transcendent and transformational miracles helps explain why many scientists, especially those focused on narrowly specialized disciplines, fail to detect God's involvement in the natural realm. The historical context of such miracles is part of the explanation. From a biblical perspective, creation miracles occurred long ago, "in the beginning," or during the six creation epochs. The text says God ceased from His work of creating on the seventh day, and His "day of rest" is ongoing in the context of this universe.[13] So, only research probing the era before human history (the sixth creation epoch ends with the creation of human beings[14]) would be expected to yield evidence of God's transcendent and transformational creation miracles.

CHARACTERISTICS OF CREATION

It's important to keep in mind that the Bible cannot contain every detail of history—including natural history. Just because Neptune isn't mentioned in the Bible, one cannot say that Neptune doesn't exist or isn't important. No book or set of books could possibly narrate and describe everything that has happened in the universe to date. And no team of researchers, however global, learned, transgenerational, and well-funded, can ever discover and record every fact of nature. As a result, a perfect and complete creation model will stay beyond reach, and some mysteries will always remain.

However, the existing biblical content does account for many significant characteristics of this present creation.

God's biblically stated purposes for the natural realm provide essential insight to many issues of dispute among and between both creationists and evolutionists. Much debate surrounds scientists' observations and measurements of the following features in particular, all of which the Bible addresses either explicitly or implicitly:

- The physical laws and constants
- Seemingly "imperfect" designs
- The origin, dimensions, and structure of the universe
- The cosmic time scale
- The progression of life from simple to complex
- Interruptions to life, including the flood of Noah's day
- The existence of common designs among creatures

In the following pages, these characteristics are examined and discussed from the perspective of a biblical creation scenario.

Physical Laws and Constants

Through the prophet Jeremiah, God says that the laws governing the heavens and Earth are "fixed."[15] God determined what they would be from the beginning, and their values align precisely with His unfolding plans. RTB's biblical creation model expects scientific research and exploration of the cosmos to show that physical laws do not vary to any significant degree throughout space and time.

Some creationists argue that Adam and Eve's rebellion against God's authority brought about major changes in physics—either the first appearance of decay, a dramatic rise in the level and/or rate of decay, or a huge decline in a hypothetical divine counterbalancing action. Neither the biblical text nor the scientific data warrant such an assertion. In fact, the scientific data flatly contradicts it.

Prior to Adam's sin, the second law of thermodynamics would have already been in operation. According to Paul's writing in Romans, the entire creation has been "groaning" from the start and right up to the present time as a consequence of its "bondage to decay."[16]

Genesis (along with other Bible passages) affirms that the physical realities of sunlight, starlight, metabolism, and human work preceded human sin.[17] The possibility of the sun's (and other stars') stable burning, organisms' metabolizing of food, and Adam's (pre-sin) performance of work would all be ruled out if the laws of physics had not been in continuous operation at their current values. Even a tiny variation would have rendered all these functions impossible.[18]

Seemingly "Imperfect" Designs

Throughout the account of the six creation days in Genesis 1, the narrator quotes God's evaluation of His creation as "good." This assessment appears six times in the passage.[19] Finally, as God surveys the whole sweep of His creative work, He is quoted as calling it "very good."[20]

Some readers, both evolutionists and creationists, interpret these words to mean "perfect." Evolutionists point out apparent imperfections in nature as evidence against biblical creation, while creationists defend perfection and blame sin for any apparent imperfection. But context does not necessitate such an interpretation.

"Very good" seems more reasonably to imply that nature manifests appropriate design for fulfillment of its purposes. The ideal design for a given purpose must not be confused with ultimate perfection. The new, truly perfect creation is said to follow this one. The current creation is optimal in the sense that it perfectly suits God's plans for it—the triumph of life over death, good over evil, love over apathy, light over darkness, and freedom over bondage.

Achievement of these victories requires the operation of thermodynamic laws, including the pervasive law of decay. The Bible explains why. The creation's "bondage to decay" plays a crucial part in God's plan to bring His followers to "glorious freedom."[21] The decay process, which includes what physicists call the second law of thermodynamics, implies that God's very good designs will degrade over time. Researchers may expect optimally designed creations to gradually accumulate imperfections, perhaps at least partly for the purpose of heightening human yearning for and anticipation of the perfect creation to come.

The Origin, Dimensions, and Structure of the Universe

Strict naturalists can explain the dimensions and structure of the universe essentially as remarkable happenstance. They can observe and measure "what is," but meaning is only what humans assign. Creation myths venture to do more. And while most make reference to "the beginning," they typically pick up *after* something already exists. In other words, they tell a story that begins after the beginning. None gives the abundance of specific, testable detail about the beginning—detail that proves accurate—presented by Bible authors, all writing prior to second century AD. By way of review, these details are among the facts confirmed by 20th-century cosmologists (see chapter 3, pages 45–49):

1. The universe has a beginning in finite time (Genesis 1:1; 2:3-4; Psalm 148:5; Isaiah 40:26; 42:5; 45:18; John 1:3; Colossians 1:15-17; Hebrews 11:3).
2. The beginning of space and time coincides with the beginning of the physical universe (Genesis 1:1; Colossians 1:15-17; 2 Timothy 1:9; Titus 1:2; Hebrews 11:3).
3. The material universe was made from that which is immaterial (Hebrews 11:3).
4. The universe has been continuously expanding from the beginning of space and time (Job 9:8; Psalm 104:2; Isaiah 40:22; 42:5; 44:24; 45:12; 48:13; 51:13; Jeremiah 10:12; 51:15; Zechariah 12:1).
5. The expansion of the universe appears precisely guided for the benefit of life (Job 9:8; Isaiah 44:24; 45:12; 48:13).
6. The expansion of the universe resembles the spreading out and setting up of a tent (Psalm 104:2; Isaiah 40:22).
7. The universe functions according to fixed physical laws (Jeremiah 33:25).
8. The entire universe is subject to those physical laws (Romans 8:20-22).
9. The universe has an ending in finite time (Job 14:12; Ecclesiastes 12:2; Isaiah 34:4; 51:6; 65:17; 66:22; Matthew 24:35; Hebrews 1:10-12; 12:27-28; 2 Peter 3:7,10-13; Revelation 21:1-5).
10. At its end, the universe will roll up like a scroll and vanish in a burst of extreme heat (Isaiah 34:4; 2 Peter 3:7,10; Revelation 6:14).

Statements 4, 7, and 8 imply that the universe must get colder as it grows older. Statement 6 may suggest that just as a tent is a surface with no physical center, so too the universe has no physical center or anything either interior or exterior to its surface. Thousands of years previous to any scientific speculation or research into big bang cosmology, the Bible predicted all of the fundamental attributes of a big bang universe. Scripture even implies, in its account of the demise of the cosmos, what kind of big bang universe this is (see "For a Limited Time Only," page 77).

Creation Time Scales
Considerable dispute surrounds the issue of cosmic age. Evolutionists see an ancient cosmos as evidence against creation. Why would God waste billions of years if all He wanted was a planet with humans? On this issue, creationists typically agree. They say He didn't need all that time and that's

why six 24-hour creation days a few thousand years ago makes sense.

Time markers for creation-week events in Scripture are more consistently interpreted as relative to one another rather than absolute. Numerous metaphors give a sense of great age—eons as opposed to just a few millennia of cosmic and geologic history. God's eternal existence, for example, is compared to the duration of the mountains, oceans, rivers, hills, soil, and fields (Psalm 90:2; Proverbs 8:24-26; Ecclesiastes 1:4-10). Micah 6:2 refers directly (rather than metaphorically) to the great antiquity of Earth's foundations. Habakkuk 3:6 declares the mountains "ancient" and the hills "age-old." Peter says the heavens and earth existed "long ago" (2 Peter 3:5).

Great age can also be inferred from biblical metaphors for the size of the cosmos—an expanse of stars said to be as numerous as Earth's grains of sand.[22] Given star formation rates, average distances between stars, and the fixed velocity of light, this number translates to an age greater than several hundred million years, at least.[23]

In light of what the Bible says about His unlimited power,[24] God could have chosen any time scale, however short or long, to perform His creative work. As for the six "days" of creation, Hebrew allows for more than one literal possibility. The word translated "day" in Genesis 1, *yôm*, has four different literal definitions: (1) a portion of the daylight hours, (2) all of the daylight portion of a 24-hour day, (3) a 24-hour day, and (4) a long but finite time period.[25] (Unlike English or modern Hebrew, biblical Hebrew had no other word for a finite era or epoch.) RTB's model posits that the fourth definition affords the greatest consistency with all the biblical creation accounts (see list on page 54). A thorough analysis of the biblical, theological, and scientific evidence for long creation days can be found in *A Matter of Days*. Chapters 7 and 8 (pages 127–162) go on to discuss why God would have taken so long in the creation process.

Progression from Simple to Complex Life

Long before any human possessed the tools to study life's history, Bible writers described *evolution* in its simplest meaning—"change with respect to time." Genesis 1 and 2, Job 38–42, and Psalm 104 all depict a series of purposeful and progressive creation events—from the origin of simpler life-forms, such as bacteria, to the more complex, such as human beings—a pattern no ancient person would be expected to know. Nor is it a pattern random, natural processes would necessarily yield.

Genesis 1:2-5 implies that the first life created on Earth was simple marine life[26] and that it appeared while Earth was still emerging from extremely hostile-to-life conditions.[27] Genesis 1, with supporting material from Job 38–42 and Psalm 104, outlines the progression, brought about by divine intervention, from primitive marine life to plant life on newly forming continents, to swarms of small sea animals, to birds and sea mammals, to three kinds of advanced land mammals, and, finally, to human beings.[28]

For a Limited Time Only

The Bible explicitly states that the universe will not last forever. The apostle Peter declared, "The heavens will disappear with a roar; the elements will be destroyed by fire. . . . That day will bring about the destruction of the heavens by fire, and the elements will melt in the heat."[29] Isaiah wrote, "All the stars of the heavens will be dissolved and the sky rolled up like a scroll."[30]

These Bible passages are metaphorical, and some scholars debate how literally one should interpret these depictions of how the universe will end. One class of big bang models, however, predicts a cosmic end that is remarkably consistent with the heat imagery.[31]

The ekpyrotic big bang models propose that instead of the universe being laid out as a 10-dimensional flat sheet, the sheet instead folds to form a gigantic "U," which means one 10-dimensional flat surface is opposite another 10-dimensional flat surface. If the two surfaces are close enough together for a long-enough time, a quantum fluctuation in the space-time fabric of one of the two surfaces could peel off and make contact with a quantum fluctuation in the space-time fabric of the other. If that contact were to happen, the space-time fabric of the entire universe would curl up and disappear, taking the whole universe with it in a fiery implosion.

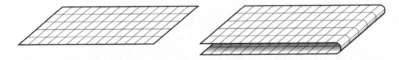

Figure 4.1: Ekpyrotic Cosmic Creation Model
The plane on the left represents the standard 10-dimensional flat universe models (only two dimensions are depicted). The folded plane on the right represents the ekpyrotic cosmic models — a 10-dimensional flat sheet folded over so that two portions of the 10-dimensional sheet exist very close to each other. If the two faces of the folded sheet sit less than a millimeter apart, a reasonable probability would exist (within the time frame of 10 to 20 billion years) for a quantum fluctuation in the space-time fabric of one to make physical contact with a quantum fluctuation in the space-time fabric of the other. If such contact were to occur, the space-time fabric of the entire folded sheet would collapse into the contact point, causing the entire universe to end in a fiery demise. Such an end to the universe would fit well with the metaphorical descriptions in Isaiah 34:4; 2 Peter 3:7,10; and Revelation 6:14.

This progression in complexity takes place on more than one level. The biblical creation accounts claim the first life-forms were purely physical in nature. Relatively late in creation history the first "soulish" animals arrived on the scene. These creatures, predominantly birds and mammals, show limited capacity for emotion, volition, and intellect. They can relate not only to each other but also to human beings, many with the ability to interpret tone of voice and respond to training.[32] The last species of life to appear on Earth, human beings (*Homo sapiens sapiens*), manifests not only far greater soulish capacities but also a unique characteristic the Bible identifies as "spirit." No other creature expresses spirituality, as defined by an innate

- Awareness of right and wrong, or conscience
- Awareness of mortality and concerns about what lies beyond death
- Hunger for hope, for a sense of purpose and destiny
- Compulsion to discover and create
- Capacity to recognize beauty, truth, logic, and absolutes
- Propensity to worship and communicate with deity (or deities)

With respect to soulish and spiritual animals, the Hebrew verbs in the relevant texts—'*āśâ, bārā'*, and *yāṣar*—indicate the direct miraculous intervention of God in the introduction of each particular type of creature. Such verbs seem to preclude the possibility that God worked through natural-process evolution to produce these special characteristics. Any viable creation and/or evolution model must account for the emergence not only of Earth's first life but also of soulish and then spiritual life.

Interruptions to Life
More than once, the Bible suggests that life on Earth was seriously disrupted, possibly even extinguished on some scale. But always it returned, each time with just-right characteristics suited to the changing environment. And all the while, Earth was gaining the optimal biomass and biodiversity to support humans. Gradually Earth built up sufficient biodeposits for the launch and rapid development of global human civilization and technology. The Creator appears to have used mass creation, extinction, and re-creation (or speciation) events to store up resources—including the layers of organic material, deposits of fossil fuels, and deposits of ores and minerals—all ideal for the benefit of humanity. Psalm 104:27-30 depicts this cycle of life and death:

> These [creatures] all look to you
>   to give them food at the proper time.

When you give it to them,
    they gather it up;
when you open your hand,
    they are satisfied with good things.
When you hide your face,
    they are terrified;
when you take away their breath,
    they die and return to the dust.
When you send your Spirit,
    they are created,
    and you renew the face of the earth.

This story of life, death, and new life must be part of a biblical model. Such a model must not be misconstrued, however, as contradictory to New Testament statements about the kind of death that originated with Adam.[33] Romans 5:12 says, "Sin entered the world through one man, and death through sin, and in this way death came to all men, because all sinned."

Two qualifications clarify that the death introduced by Adam's sin applies strictly to humans.[34] First, the whole of Scripture confirms that only humans, among all life created on Earth, can (and do) sin. Therefore, this "death through sin" would apply to humans alone, not to plants and not to other animals. Second, the passage states specifically that this "death came to all men." It does not say "to all creation" or "to all creatures." It makes no apparent reference to plant or animal life, nor do other parallel passages.[35]

One major interruption to life specifically described in Scripture would be the flood of Noah's day. It occurred at a time when humanity was still localized and, according to Genesis 6:5, had become dangerously corrupt. For the sake of humanity's long-term survivability, God sent a flood to cleanse the earth of all but one family of humans and all but a few of the soulish animals associated with humanity. This flood was, by far, more catastrophic to humans than any other, before or since, in that it so nearly extinguished the human species. But the biblical text cannot be correctly construed as lumping all Earth's cataclysms into this one event. The Bible does not declare it the source of all Earth's major geologic/tectonic features, as some creationists argue. In fact, careful analysis of the relevant biblical texts shows that Noah's Flood was geographically limited (see appendix E, "The Purpose and Extent of Noah's Flood," pages 223–225).[36]

Common Designs Among Creatures
Unlike standard evolutionary models, a biblical creation model predicts common designs even among diverse species. Many Bible passages,

particularly Job 38–41 and Psalms 104 and 139, say that when God creates He uses an optimal design. Therefore, because what works well for one species will likely be best for one or more other creatures, a biblical creation model anticipates that scientists should discover many examples of shared designs—or common morphology.

An optimally designed organ, limb, appendage, or vessel will serve a wide variety of creatures. For example, the lung has an unsurpassed, even unsurpassable,[37] respiratory efficiency for large-bodied air-breathing animals. Therefore, a biblical perspective anticipates that all large air-breathing animal species would use lungs for respiration.

Pervasive common morphological designs demand pervasive common biochemical designs. Because morphological features are specified and programmed by DNA, DNA similarities should be widespread. These similarities would be most pronounced in the framework of a biblical creation model. One God—the Bible's wise and loving Creator—would likely use the same DNA blueprints for optimized designs over and over again. Naturalistic models for life, based on chance or random outcomes, would predict a wide range of DNA diversity (see chapter 7, pages 141–146).

THE LATITUDE OF BIBLICAL LANGUAGE

In some cases, Bible authors use nonspecific language to describe God's creative activity. For example, the Hebrew verb *hāyâ* expresses how light first appeared on Earth's surface (Genesis 1:3). It asserts establishment of the water cycle (Genesis 1:6), emergence of continental landmasses (Genesis 1:9), and the first appearance of the sun, moon, and stars to serve as watchful eyes on Earth's surface (Genesis 1:14). The Hebrew verb *dāshā'* depicts the production of plants on the continents (Genesis 1:11).

Both *hāyâ* and *dāshā'* allow for some flexibility in the "how" of these phenomena. Their range of use encompasses the possibility of either a transformational miracle or a set of well-timed sustaining miracles or some combination of the two.[38] In other words, the plants and swarms of small sea creatures could have come about as the result of (1) God's direct, momentary, miraculous intervention, (2) God's guidance and timing of natural processes, or (3) the two together in any combination. Therefore, the debate over exactly how plants and throngs of small sea life originated cannot be settled by the biblical data alone. In this instance, the record of nature potentially holds greater detail.

For other events, the text gives a clear indication of causality. The use of a more specific verb, *bārā'*, indicates that the universe came about through a transcendent miracle. The use of two Hebrew verbs, *bārā'* and *'āśâ*, for soulish animal and human life, implies that these creatures came into existence through a combination of transcendent and transformational miracles. Genesis and other biblical texts also seem to place some boundaries around speciation, saying certain creatures reproduce "according to their kinds."[39]

GOD'S CREATIVE OPTIONS

The Bible ascribes both *immanence* and *transcendence* to the Creator. God is said to fill the entirety of His creation, His presence permeating the whole of it. And yet the universe does not and cannot contain Him. God's transcendence means He exists in complete independence from matter and energy, the laws and constants of physics, and the space-time dimensions of the cosmos. God's freedom to operate and create knows no boundaries.

So God *could* have created a very different world. In fact, the closing chapters of the New Testament support the idea that God will someday do that very thing. As described in Revelation 21 and 22, the new creation apparently will operate by radically different physics and radically different dimensionality from the current creation (see table 4.1, below). Characteristics of the universe, of Earth, and of life evidently reflect God's choice and design. RTB's model of creation, therefore, expects that God's purposes give meaning to every observed property of the universe and of physical life. Connections between revealed purposes and discernible properties, including the laws of physics and the features of physical life, should become clearer as understanding of either (or both) increases.

TABLE 4.1: LIFE IN THE NEW CREATION

Some of the characteristics of the new creation, as described in Revelation 21–22 and elsewhere in the Bible, include these:[40]

- A more expansive habitat for humanity than what is possible in this universe
- Radically different creation laws and constants (no thermodynamics, no gravity, no electromagnetism)
- No decay, no death, no pain, no evil, no regrets, no grief
- No darkness, no shadows
- No sun, no stars, and yet light everywhere
- Different dimensionality
- Unimaginable splendor, joy, beauty, peace, and love
- No marriage (or sex), no families because multiple, simultaneously intimate relationships will eclipse the need or desire for them

- Unlimited relational delight
- Unlimited capacity for pleasure
- Wholly meaningful and satisfying work
- Opportunity to lead and instruct angels

In the words of the apostle Paul, no human in the current creation can possibly conceive or imagine how glorious life in the new creation will be.[41]

ANTICIPATION OF DISCOVERY

To a 21st-century reader, the connections between the biblical and scientific data as set forth in these pages may seem too obvious. In other words, they may appear as "hindsight" links—as if the interpreter worked backward from current knowledge to find texts that fit. One may wonder if perhaps manuscripts were altered through the years to fit emerging discoveries. This challenge deserves reflection and a response.

These three considerations seem most relevant and helpful:

- Complete New Testament manuscripts date back more than 1,700 years. Portions date back more than 1,900 years.
- The most recently written Old Testament creation account, Isaiah 40–51, dates back more than 2,700 years. One of the Dead Sea Scrolls dating to the second century BC contains the entire book of Isaiah. Its content is essentially identical to that in much later manuscripts. Such an early-dated manuscript reasonably eliminates all possible authors except the Isaiah of King Hezekiah's court, and it reasonably eliminates any significant tampering with the text.
- Only in the past few centuries have people become aware that (a) the history of life on Earth progresses from simple to complex, (b) the laws of physics are fixed and apply to the entire universe, (c) the universe is continuously expanding, (d) Earth once had no continents, (e) conditions on early Earth were extremely hostile to life, and so on.

The list for modern discoveries forecast in Scripture could go on to include many facts about stars, animal husbandry, farming, medicine, sanitation, meteorology, tectonics, and many, many more. In the second century BC (or even in the New Testament era), the most learned scholars had little or no knowledge of these facts about the natural realm—outside of what the Hebrew Scriptures recorded.

TABLE 4.2:
SUMMARY OF BIBLICAL MATERIAL ON ORIGINS AND NATURAL HISTORY

THE UNIVERSE
- Began (once) in finite time
- Coincides, at its beginning, with the beginning of space and time
- Is made from that which is immaterial
- Continuously expands from the beginning
- Is governed by constant laws of physics
- Manifests precise fine-tuning for humanity's benefit
- Has enormous volume, encompassing an "uncountable" (to ancient peoples) number of stars
- Contains stars that differ from one another
- Will someday cease to exist, though people will not

EARTH
- Emerged from the cosmos at a specific time
- Was enshrouded at its beginning by an opaque cloud layer
- Began with an ocean that covered the whole surface
- Was/is precisely fine-tuned for humanity's benefit
- Contains the resources essential for launching and sustaining human civilization
- Has a sun and moon and other astronomical companions specially designed to benefit life and humanity
- Carries finite resources and time-limiting conditions for sustaining human civilization

LIFE
- Began early in Earth's history
- Began under hostile conditions
- Began by divine intervention
- Appeared in abundance, diversity, and for long eras for the specific benefit of humanity
- Began with optimal ecological relationships
- Began with optimal design for environmental conditions
- Reflects shared common designs
- Progresses from simple to complex through a series of extinction and miraculous replacement (speciation) events
- Started as physical only (most life-forms), then came soulish creatures (many species), and finally, then came one spiritual species, an original pair of humans and all their descendants

HUMANITY

- Arrived late in Earth's history
- Represents the culmination of God's creation work on Earth
- Resulted from divine intervention
- Remains the only earthly creature with a spiritual nature
- Descended from one man and one woman who lived in a God-designed garden near the juncture of Africa, Asia, and Europe
- Migrated rapidly from area of origin shortly after the flood of Noah's time
- Experienced significant drop in potential life span after the time of that flood
- Genetically bottlenecked at a later date for males than for females (because male Flood survivors were biologically related)
- Was gifted from the outset with attributes needed for functioning in a high-tech civilization

THE SCIENCE CHALLENGE

From the biblical texts a framework emerges, one that explains the various features of the natural realm in logically and scientifically testable terms. (For a summary of biblically described and anticipated features, see table 4.2, pages 83–84.) The RTB creation model, based on this framework, invites side-by-side comparison with competing models on the scientific findings already established and still accumulating. Whole books explain (in great detail) various aspects of RTB's testable/falsifiable, predictive model.[42] Future books will offer more adequate treatment of several other specific creation/evolution issues. For now, the following chapters briefly highlight some of the ways RTB's creation-model-in-progress fares in light of well-established evidence. The most recent scientific discoveries seriously test the mettle of the Bible's creation story.

CHAPTER 5

THE COSMOS TESTS THE RTB CREATION MODEL

News flash! The Martians have landed! A capsule creaks open. A hissing, tentacled creature emerges, seemingly fascinating and horrifying the newscaster. Death and destruction follow—or do they? Within minutes for some, a few hours for others, reports from various witnesses under much broader observing conditions revealed the correct explanation. The confusion cleared fairly quickly.

Compiling data about cosmic origins has taken far longer. But the long wait has produced amazing results. Probably no discipline of science has seen such remarkable recent advances in both observation and theory as cosmology. At the beginning of the 20th century, astronomers had seen only a tiny fraction—less than one-millionth—of the potentially observable universe. Today, they can see all of it, all the way to the very limits imposed by the laws of physics. In 1900, astronomers lacked a clear picture even of the Milky Way Galaxy and its spiral structure. Today, they have mapped out virtually all of its visible matter and most of its dark matter. In 1900, astronomers had not yet established the existence of other galaxies. Today, they can report that the observable universe contains about 200 to 300 billion galaxies. They can also say approximately how many of each type of galaxy (for example, spirals, ellipticals, irregulars, quasars, and Seyferts) the universe contains. In 1900, astronomers had no idea how much of cosmic history they were observing. Today, astronomers can say with confidence they have witnessed all of cosmic history.

Astronomers use both Earth-based and space-based telescopes to look back in time across the entire electromagnetic spectrum to observe the total sweep of cosmic history from its beginning 13.73 billion years ago to the present moment. As they aim their telescopes at progressively more distant objects, they directly observe the past. The velocity of light serves as a time machine. Since light takes a finite time to travel a certain distance, scientists can directly map the state of the universe at any epoch they desire simply by observing an object at the appropriate distance. This enterprise is like leafing through a detailed photo album of a middle-aged person whose life is documented from the time he or she formed in the mother's womb until the latest photo was taken.

Armed with their telescopic time machines, astronomers today can put scientific models for the origin, structure, and history of the universe to ever more rigorous tests. The viability of RTB's cosmic creation model can be evaluated in light of these discoveries. While the model's development remains an ongoing process, its foundations and scientific assessments are explained more fully in *The Creator and the Cosmos* and *Beyond the Cosmos*.[1] A brief summary appears in "Cosmic Creation Highlights" (see page 87). Comment on some distinctive features of the model follows the summary, along with specific scientific evaluations.

Figure 5.1: Extent of Cosmic History Astronomers Directly Observe
If the 13.73-billion-year history of the universe is represented by a 24-hour clock, the portion of cosmic history astronomers *directly* observe is equivalent to 23 hours 59 minutes 58 seconds. That is, astronomers directly observe all but 0.003 percent of the entire history of the universe. The latest release of data from The Wilkinson Microwave Anisotropy Probe (WMAP) satellite gives astronomers an additional *indirect* view, so that the new total view would equate to 23 hours 59 minutes 59.999999999999999999999999999999 999999999999999 seconds.[2]

Cosmic Creation Highlights

The RTB cosmic creation model is based on the following biblical premises:

- The universe has a beginning, a single beginning as opposed to multiple beginnings. This beginning was the start of matter, energy, space, and time. The beginning points to a Beginner who exists beyond the laws of physics and the space-time dimensions of the universe.
- The universe operates under fixed laws of physics.
- The universe is enormous, containing what for ancient peoples would have been an uncountable number of stars.
- The universe has been extraordinarily fine-tuned or designed to make the existence of life possible. This cosmic fine-tuning becomes increasingly more extraordinary in the progression from bacterial life's requirements to the necessities for human life.
- The universe manifests the most significant features of the big bang model as predicted by the Bible thousands of years ago.
- Humanity has been placed at the best cosmic location and at the best cosmic time to observe the beauty, power, wisdom, and love of the Creator in creating and shaping the cosmos for their benefit.

Figure 5.2: Detailed Map of the Cosmic Background Radiation
The WMAP has delivered the most detailed map ever produced of the radiation left over from the cosmic origin event. The variations in shading show temperature variations from region to region in the early universe. The Sloan Digital Sky Survey of hundreds of thousands of galaxies established that the warmer regions in the cosmic creation event were the "seeds" that grew into galaxies and galaxy clusters. The above map is from the second data release and is based on three years' continuous observation. Each circular image represents one hemisphere of the all-sky survey.

(Photo courtesy of WMAP Science Team)

A Cosmic Beginning

Several independent lines of measurement have now determined the birth date of the universe to be 13.73 billion years ago.[3] Given that the universe contains mass and the equations of general relativity reliably describe the dynamics of the universe, the space-time theorems of general relativity establish that the space-time dimensions associated with the universe must have had a specific beginning. Within the universe, time is the dimension in which cause-and-effect relationships occur. Effects follow their causes. So the beginning of cosmic time implies that an Agent (cause) outside the universe's space-time dimensions is responsible for bringing into existence the space, time, matter, and energy (effects) astronomers observe.[4]

By 1970 astronomers had developed only one space-time theorem of general relativity, one that applied within the framework of classical general relativity. Today, a whole family of space-time theorems exists, and these are applicable for all the inflationary cosmic models (models wherein the effect of general relativity is augmented by a "scalar field" that expanded the universe at many times the velocity of light during a brief period when the universe was younger than a quadrillionth of a quadrillionth of a second), as well as for all reasonable quantum gravity models.[5]

Theoretical physicists have further demonstrated that all other cosmic models (those competing against general relativity) would either produce an unstable universe or demand conditions contradicted by well-established observations. Two such physicists agree that all reasonable cosmic models (models that do not demand a violation of the thermodynamic laws) must be subject to "the relentless grip of the space-time theorems."[6]

Cosmologists don't doubt that the universe contains mass. Neither do most people. However, at the time the first space-time theorem of general relativity was published, astronomers had determined to only 1 percent precision that general relativity reliably describes the dynamics of the universe. At the time, they had performed only three independent tests of general relativity.

Today, the reliability of general relativity in describing the dynamics of the universe has been confirmed to 0.000000000001 percent precision, and astronomers have performed more than a dozen independent tests of general relativity. In the words of the British mathematical physicist Sir Roger Penrose, coproducer (with Stephen Hawking) of the first space-time theorem, "This makes Einstein's general relativity, in this particular sense, the most accurately tested theory known to science."[7] The thoroughness of testing and precision of results leave no reasonable basis for doubting that a causal Agent outside space and time brought the universe of space, time, matter, and energy into existence.[8]

(Photo courtesy of James King-Holmes/
Photo Researchers, Inc.)

(Photo courtesy of Anthony Howarth/
Photo Researchers, Inc.)

Figure 5.3: Stephen Hawking and Roger Penrose
In 1970 British physicists Stephen Hawking and Roger Penrose published the first of many space-time theorems springing from general relativity. Their theorem established that both space and time began at the cosmic beginning — *if* general relativity proved reliable. At the time, some astronomers still held to a small uncertainty about the theory's accuracy in describing cosmic dynamics. By the beginning of the 21st century, however, overwhelming observational evidence had eliminated that uncertainty.

More than a century of progress in cosmology contradicts the fundamental belief of philosophical naturalism—the claim that all causes and effects are contained within nature. Theoretical and observational evidences for a transcendent cosmic origin, on the other hand, coincide with biblical descriptions in at least two ways: (1) The Bible claims a cosmic beginning, and (2) the biblical God transcends the universe. RTB's biblical creation model anticipates the continued accumulation of evidence for ongoing refinement of big bang (expanding universe) models. It also expects evidence to grow for the transcendent causal Agent implied by the space-time theorems of general relativity.

When the Bible describes God's triune intra-relationship (the Father, Son, and Holy Spirit relating to one another) and causal activity (for example, putting His "grace" into effect) as being outside the limits of space-time, matter, and energy,[9] it suggests the possibility of some kind of dimensionality and/or temporality beyond the universe. The declaration in Hebrews that the detectable universe was made from that which cannot be detected implies the same possibility.[10] Such a possibility remains consistent with the singularity feature of most (though not all) big bang models. Time and further research will test the accuracy of this particular feature of the RTB creation model.

EXTRA DIMENSIONS

While mathematicians can prove theoretically that in four independent dimensions of space, a three-dimensional basketball can be turned inside out without making a cut or a hole in its surface, limitations on human imagination prevent people from being able to visualize or picture exactly how such a phenomenon can occur. People simply cannot visualize phenomena in more dimensions than they personally experience.

This limitation has both theological and scientific implications. Religions that are mere human inventions (lacking inspiration from a Being beyond the dimensions of length, width, height, and time) will be devoid of transcendent (yet consistent) teachings. They won't include information that demands the existence of entities, phenomena, and/or dimensions beyond the familiar four. (Though other religions might appeal to magic or appeal to seemingly trans-dimensional entities, the attributes, activities, and behaviors described can be comprehended and anticipated within the dimensions of length, width, height, and time.)

The Bible, by contrast, abounds with extra- or trans-dimensional doctrines—the triune nature of God (three Persons yet just one Essence), His simultaneous transcendence and immanence, and the simultaneity of human free choice and divine predetermination, to mention a few. Scripture also contains accounts of trans-dimensional or extra-dimensional events, such as Jesus' transfiguration[11] and, after the Resurrection, His passing bodily into a locked room.[12]

The Bible thus demonstrates that at least some of its message comes from beyond length, width, height, and time. A God with the capacity to operate in or beyond extra dimensions explains how such "impossibilities" could be possible.[13]

The scientific application of this human dimensional limitation implies that if a transcendent Creator were responsible for bringing the universe into existence, He has the capacity to structure the universe with more dimensions than humans can visualize. Because the Bible declares that the Creator intends His physical creation to reveal both His existence and His nature to humanity,[14] the possibility that He did so *could* be discoverable.

The likelihood that the universe was constructed with more dimensions than the familiar length, width, height, and time has recently received dramatic verification. Breakthroughs came as physicists and astronomers tackled two seemingly intractable problems plaguing the big bang creation models.

· The first dilemma was that treating fundamental particles as point entities (the traditional view) made unification of any of the four fundamental forces of physics (gravity, electromagnetism, and the strong and weak nuclear forces) impossible. Complete theoretical and experimental proof that this unification can and did occur for the weak nuclear force (the force governing radioactivity) and the electromagnetic force[15] made a new approach necessary, one that allowed more flexibility. Lines or loops of energy called "strings" provided that new explanation.

When theoreticians treated fundamental particles as highly stretched, vibrating, rotating "elastic bands" in the extreme heat of the first split second after creation, the dilemma resolved itself. For all practical purposes, fundamental particles have behaved as points under the cooler conditions since that crucial early moment, but not in the initial instant. These strings, however, require the existence of more than three spatial dimensions. They need more room to operate.

In the easily recognized four space-time dimensions of the universe, all gravitational theories contradict the possibility of quantum mechanics, and all quantum mechanical theories contradict the possibility of gravity. This head-on collision posed the second dilemma. Andrew Strominger hypothesized a brilliant solution in the form of "extremal" (very small) black holes, which become massless at critical moments.[16] At first, however, he seemed merely to have traded one dilemma for another. Black holes are massive objects so highly collapsed that their gravity attracts anything nearby. How could a black hole be massless without violating the definition of a black hole or without violating the principles of gravity? Simply put, how can there be gravity without mass?

Extra-dimensionality once again supplied the answer. Strominger discovered that in six spatial dimensions, the mass of an extremal black hole is proportional to its surface area. As the black hole's surface area shrinks, the mass eventually drops to zero. The possibility of zero-mass black holes leads to the possibility that gravity and quantum mechanics can fully coexist. However, this complete coexistence demands a total of nine space dimensions (the familiar length, width, and height, plus six more).

One theory solves two great dilemmas: The universe was created with 10 initially rapidly expanding space-time dimensions. When the universe was just 10^{-43} seconds old (a 10-trillionth of a quadrillionth of a quadrillionth of a second), gravity separated from the strong-electroweak force, and at that moment six of these 10 dimensions ceased to expand. Today, these six dimensions still exist as components of the universe, but they remain as tightly curled up as when the cosmos was only 10^{-43} seconds old.

Six sets of evidence indicate that this theory is correct.[17] Perhaps the most convincing is that string theory produces, as a bonus byproduct, all the equations of special and general relativity. In other words, if scientists knew nothing at all about relativity, this 10-dimensional string theory would have revealed relativity theory in complete form. Therefore, the experimental confirmation of special and general relativity establishes (to the same remarkable degree of certainty) that 10 space-time dimensions somehow frame the physical universe. (Exactly what form or shape these 10 dimensions take or whether or not an 11th dimension somehow interacts with the 10 remains to be determined.)

The scientific discoveries and evidences that establish the existence of 10 dimensions (one of time, nine of space) for the universe and that imply (from the space-time theorems of general relativity) that the cosmic Creator (causal Agent) operates in or beyond at least the equivalent of one additional time dimension provide important confirmations of RTB's biblical creation model. *Beyond the Cosmos* recounts in greater detail the discovery of extra dimensions and how that discovery attests and elucidates various aspects of the Christian faith.

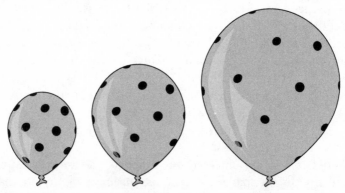

Figure 5.4: An Analogy for Cosmic Expansion
Like dots on the surface of a balloon, all the matter and energy of the universe (stars, planets, galaxies, and so on) exist on its surface. Unlike the balloon's surface, which is two-dimensional, the cosmic surface is three-dimensional (if one ignores the six extremely tiny space dimensions that apparently remain tightly curled). As the dots on an expanding balloon move farther (and faster) away from each other while the balloon expands, so do the galaxies on the cosmic space surface move farther (and faster) apart as the universe expands. However, whereas the balloon dots themselves grow bigger with the balloon's expansion, local gravity from the mass within each galaxy prevents the galaxies from expanding significantly as the universe expands.

CONTINUAL COSMIC EXPANSION

The continual expansion of the universe from the creation event onward is the most frequently described feature of the cosmos found in the Bible.[18] This biblical claim predates by more than 3,000 years the scientific discoveries that verify it. The Bible specifically states that God "alone stretches out the heavens."[19]

The first observational evidence of this phenomenon came from a set of measurements that established the "law of redshifts." According to this law, the velocities at which galaxies move away from Earth are directly proportional to their distances. A classic analogy illustrates the law. As dots painted on the surface of a balloon continuously expand while the balloon is being blown up, the dots also move away from one another at rates directly proportional to the distances separating them (see figure 5.4, page 92).

Photo-images demonstrating that extremely distant galaxies are jammed much more tightly together than are galaxies nearby supply the most visually obvious evidence for continual cosmic expansion.[20] Because it takes time for light from distant sources to travel to the astronomer's telescope, the farther away astronomers look, the farther back in time they see. Therefore, the observation of galaxies closer and closer together as astronomers look farther and farther away confirms that the universe has been continuously expanding since the cosmic beginning.

Surface brightness measurements of various astronomical bodies also demonstrate that the universe has been, and still is, expanding.[21] The surface brightness (luminosity divided by the radius squared) of identical objects in a nonexpanding universe would greatly exceed that of identical objects in an expanding universe, due to its increasing area. The measurements unambiguously point to expansion.

If the universe indeed has been continuously expanding for billions of years, the most distantly observed objects would be moving away at extremely high velocities (speeds close to the velocity of light). Einstein's relativity theory predicts that these distant "clocks" will thus run about 10 to 60 percent more slowly (due to relativistic effects) than equivalent clocks in the vicinity of the Milky Way Galaxy. Observations show that in the Milky Way Galaxy, an exploding star (a supernova) takes about seven months to transition from maximum to minimum brightness, and a typical gamma-ray burst takes an average of about 15 seconds to undergo this same transition. However, at great distances these transitions take longer by the exact amounts consistent with their distance in an expanding universe—one that's been continuously expanding for the past 13.73 billion years.[22]

The existence of planets and solar-type stars is possible only if the universe continuously expands—and does so at a just-right rate for a just-right duration. A universe that expands too slowly produces only neutron stars and black holes.[23] A universe that expands too rapidly produces no stars at all and thus no planets.[24]

Astronomers and physicists note that the two factors governing the cosmic expansion rate reflect the most exquisite fine-tuning noted anywhere in the sciences. According to recent studies, for the universe to produce the kinds of galaxies, stars, planets, and chemical elements essential for the existence of physical life, the cosmic mass density must be fine-tuned to at least one part in 10^{60}. The cosmic dark energy density (the self-stretching property of the space surface) on which all the matter and energy of the universe resides must be fine-tuned to at least one part in 10^{120}.

To put this number into perspective, it exceeds the number of protons and neutrons in the observable universe by 100 billion quadrillion quadrillion times.[25]

In the face of such staggering fine-tuning, even nontheistic scientists have made bold concessions. One research team said, "Arranging the universe as we think it is arranged would have required a miracle. . . . It seems an external agent intervened in cosmic history for reasons of its own."[26]

On the Cosmic Surface

Three thousand years ago, a psalmist depicted God's stretching out the universe as the unfolding of a tent.[27] Isaiah used the same imagery 2,700 years ago, saying God stretches out the heavens as one would stretch out a tent to live in it.[28] These biblical word pictures for the universe imply that, like a tent, the physical cosmos is a surface. Such a concept for the universe stood radically apart from the cosmologies of the other religions, philosophies, and sciences of the pre-20th-century world.

Today, the biblical concept of a cosmic surface can be put to the test. The expanding balloon analogy mentioned earlier not only illustrates the continual expansion of the universe, but it also helps visualize its geometry. The balloon's physical material is its two-dimensional surface. No balloon stuff resides either interior to or exterior to that surface. Consequently, no item located on the balloon can be said to reside at the balloon's center.

Similarly, astronomers observe from the positions and dynamics of the universe's galaxies that all the matter and energy of the universe reside on the cosmic surface. According to general relativity, all the space-time dimensions are likewise confined to the cosmic surface. So, too (as with the balloon), no star or galaxy can be said to reside at the cosmic center. It is literally impossible for any physical entity to exist at the "center" of the space-time surface.

The balloon analogy obviously breaks down in its shape and number of dimensions. With respect to the large dimensions of the universe, the surface is three-dimensional, not two. Instead of a spherically shaped surface, the cosmic surface is nearly flat in its geometry.

An Awful Waste of Space?

The extravagance of the universe has mystified people for centuries. The movie *Contact*, based on Carl Sagan's novel, glamorized the notion that "so much space is an awful waste if life resides only on Earth." So many planets, stars, galaxies, and space all seem unnecessary if only one planet supports life. Britain's famed physicist Stephen Hawking convinced many of his devotees that it's "very hard to believe" God would make so many "useless" stars if His intent were to make a home for only one particular physical and intelligent species.[29]

Recent discoveries, however, provide at least two life-essential reasons for the enormity of the universe. First, the density of the protons and neutrons in the universe (cosmic baryon density) must be fine-tuned to support the nuclear fusion that in turn produces life's required elements. With a slightly lower baryon density (producing fewer than about 10 billion trillion observable stars), nuclear fusion would be less productive and the cosmos would be incapable of generating elements heavier than helium. Or, if the baryon density were slightly higher (producing more than about 10 billion trillion observable stars), nuclear fusion would be more productive and all the elements would quickly become as heavy as, or heavier than, iron. Either way, life-essential elements such as carbon, nitrogen, oxygen, phosphorous, and potassium would not exist.

Second, the total cosmic mass density (density of baryons plus density of exotic mass particles in the universe) plays a key role in determining the cosmic expansion rate. If the cosmic mass density were very slightly lower, gravity would be too weak to apply much braking to the cosmic expansion. The universe would expand so rapidly that matter couldn't coalesce into galaxies, stars, and planets. Life would have no home.

With a cosmic mass density very slightly greater, gravity's powerful grip would soon have collapsed all the mass of the universe into black holes and neutron stars. With a minimum density (inside or on black holes and neutron stars) of 5 billion tons per teaspoonful, such a universe would not permit atoms to exist, much less allow the existence or the assembly of life molecules.

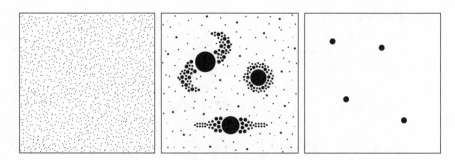

Figure 5.5: Fine-Tuning of Cosmic Expansion
A too-rapid cosmic expansion (left) would disperse matter so quickly that gravity would have no chance to concentrate gas to produce stars and galaxies. A too-slow cosmic expansion (right) would allow gravity to work so effectively as to draw all the matter in the universe into super-dense objects, such as black holes and neutron stars. However, the universe expands at the just-right rate (middle) to permit formation of the wide range of galaxies and stars necessary for the existence of a life-support planet.

These findings suggest the Creator's intentionality. They demonstrate fine-tuning for the possible support of life and humanity (one part in 10^{60}) almost as astounding as that manifested by the dark energy density. They testify of a Creator who cares enough to invest 13.73 billion years of time and 10 billion trillion stars of matter and energy in preparing a just-right home for life, including human life, even if life is only temporary. They also verify the Bible's bold statement made more than 3,400 years ago that the stars in the universe add up to what was then an uncountable number (more than a few billion).[30]

A COSMOS BUILT FOR HUMANITY

In 1961 Princeton physicist Robert Dicke noted that the universe couldn't contain physical life if any one of several constants of physics differed in value by even a slight amount.[31] His discovery led to development of the anthropic principle—the conclusion that the universe, Milky Way Galaxy, solar system, and Earth are all exquisitely fine-tuned, or engineered, so that human life can exist and flourish.

By early 2005, 77 characteristics of the universe had been identified as needing to fall within narrow ranges for any conceivable kind of physical life to exist at any time in the history of the universe.[32] In several cases, besides the amazingly narrow confines (for life) of the cosmic dark energy density and the cosmic mass density, the required degree of fine-tuning vastly exceeds the very best levels of design achieved by humans.

Astronomers are able to measure the Milky Way Galaxy and the solar system much more precisely than they can the universe as a whole. And when they do, they find even more evidence of supernatural design. Over 300 characteristics of the Milky Way Galaxy and the solar system require exquisite fine-tuning to make advanced life possible. These measurements demonstrate that even if the universe contains as many planets as it does stars, the possibility for the existence of just one planet or moon with the required conditions for advanced life falls below 1 in 10^{282}.[33]

To state the situation another way, the probability of finding a body anywhere in the observable universe with the necessary characteristics to support advanced life—without invoking divine miraculous intervention—is 10^{200} times more remote than the possibility that someone blind-folded could pick out on the first try a single marked dime that has been randomly shuffled into a pile of dimes that fills the universe.

Some scientists openly grapple with this stunning indicator of super-natural design. Physicist Freeman Dyson comments, "The more I examine the universe and study the details of its architecture, the more evidence I find that the universe in some sense must have known that we were coming."[34] Others, including astrophysicist Paul Davies, concede that all the evidence points to a personal Designer: "It seems as though somebody has fine-tuned nature's numbers to make the Universe. . . . The impression of design is overwhelming."[35]

Thus, the anthropic principle provides an efficient tool for putting competing creation/evolution models to the test. For example, the RTB creation model predicts that as astronomers learn more about the universe, the Milky Way Galaxy, and the solar system, the evidence for fine-tuning should increase. Nontheistic models on the other hand predict that as astronomers learn more, the evidence for fine-tuning should decrease (see "Objective Cosmic Tests for a Creator," page 98).

What has been the trend thus far? For the past 45 years, as astronomers have learned more about the universe, the Milky Way Galaxy, Earth's solar system, and Earth, evidence for the anthropic principle has multiplied. With respect to RTB's creation model, in the decade over which RTB has tracked the accumulating evidence, new discoveries have continued to augment at a breathtaking rate the plausibility of the biblical framework—including the Creator's powers, plans, and apparent purposes for the cosmos.[36]

A SURPRISING ANTHROPIC PRINCIPLE INEQUALITY

Brandon Carter, the British mathematician who first used the term "anthropic principle" in scientific literature,[37] observed a stunning temporal

---

Objective Cosmic Tests for a Creator

No Creator	Creator
The number of cosmic and solar system characteristics known to require fine-tuning for advanced life's existence will decrease as astronomical research advances.	The number of cosmic and solar system characteristics known to require fine-tuning for advanced life's existence will increase as astronomical research advances.
The degree of fine-tuning apparently necessary for advanced life will decline as astronomical research advances.	The degree of fine-tuning apparently necessary for advanced life will increase as astronomical research advances.
Evidence that the biblical Creator is the Designer of the universe and solar system for the benefit of humanity will become progressively weaker.	Evidence that the biblical Creator is the Designer of the universe and solar system for the benefit of humanity will become progressively stronger.

---

imbalance: the universe took billions of years to prepare for a species that has the potential to survive (according to the most optimistic estimates) no more than just a few million years. Carter calls this imbalance between the minimum possible time required for emergence of human life and the maximum time span for human survival "the anthropic principle inequality."[38]

Physicists John Barrow and Frank Tipler show that the inequality may be far more extreme than Carter imagined. They calculate that human civilization (something more than mere Stone Age subsistence survival), with the benefits of some level of technology and social organization, can last no longer than 41,000 years.[39] In the face of such a gross inequality, some researchers have hypothesized that the human species represents an anomaly—an exception to the cosmic rule—a late bloomer or a more fragile species among many possible intelligent life-forms that exist elsewhere.

Carter, Barrow, and Tipler counter with evidence that the inequality exists for any conceivable physical, intelligent species under any realistically possible life-support conditions.[40] Their conclusion rests on these numbers: 10 billion years represents the minimum time required for the formation of a stable planetary system, one with the right chemical and physical conditions for any life-form (let alone advanced life). Another 4 billion years represents the minimum time required for a planet in that system to accumulate adequate biomass and biodiversity to support the activities that define civilization. The convergence of "just-right" conditions for an advanced species to thrive and civilize in so short a time as 13.73 billion years reflects extraordinary, even miraculous, efficiency.[41]

Researchers also point out that the astrophysical, geophysical, and biological conditions necessary to sustain an intelligent civilized species do not last indefinitely. They are subject to disruption. The sun, like all other hydrogen-burning stars, continues to brighten during its most stable burning phase (see "Faint Sun Paradox," chapter 7, pages 129–133). At the same time, Earth's rotation period lengthens, plate-tectonic activity declines, and atmospheric composition varies. Earth can retain its advanced life-support capability for no more than an additional 10 million years, at best,[42] and the time window for civilization-support is narrower. Any similar planet would experience similar life-challenging changes.

Solar instability can impact civilization severely. Although the sun has burned with extraordinary stability for the past 50,000 years, this stability can last only about another 50,000 years.[43] A nearby supernova eruption, a disturbance of Earth's benign climate, an asteroid collision, a social upheaval, an environmental disaster, mass e1xtinction of one or more supporting species, a declining birth rate, or the accumulation of negative genetic mutations could easily send humanity back to the Stone Age within 20,000 years or less.[44] What's more, these events could lead to human extinction.

The numbers underscore how extreme the inequality may be. They show that the maximum survival time for advanced physical life equals about one-millionth the minimum time required for development of its survival necessities. Civilization time is briefer still. Such a ratio compels contemplation.

For the Creator to invest so much and for so long in a creation that can last only for such a brief time period speaks of both high value and high purpose.[45] Human civilization appears to exist "on purpose," with an enormous significance that extends beyond the limits of the cosmos. Further, that purpose apparently will be fulfilled in a relatively brief time frame. Rapid fulfillment of a profoundly significant purpose for humanity is a scenario straight from the pages of the Bible.

CONSTANT PHYSICAL LAWS

About 2,600 years ago the prophet Jeremiah described the "fixed laws of heaven and earth."[46] (See chapter 4, pages 73–74, for other biblical references to constant physical laws.) The "heavens and earth" phrase is the biblical Hebrew reference to the entire physical universe. Today, this biblical claim for the constancy of the laws of physics can be put to the test.

Physicists and astronomers recognize, as part of the anthropic principle, that physical life requires a universe that is exceptionally uniform and homogeneous. As expected, astronomers see the same abundance of elements and density of matter and energy no matter where they look in the universe. Life also requires that virtually all of the constants of physics remain fixed to an extremely high degree throughout all of cosmic history, or at least since the formation of protons and neutrons.

Thus far, astronomers have been able to make direct measurements of several constants' invariance throughout cosmic history. Their fixity is remarkable, in some cases no more than two parts per 10 trillion per year—at most—over the past 12 billion years.[47] Laboratory measurements by physicists yield variation limits as small as two parts per quadrillion (as a maximum) per year.[48]

CONTINUAL COSMIC COOLING

The biblical declarations of fixed laws of physics and continual cosmic expansion imply yet another important feature that can be tested. A universe that continuously expands under constant laws of physics must get colder and colder as it gets older and older.

The radiation left over from the cosmic creation event (aka the cosmic background radiation), when studied at great distances, and, thus, farther back in time, measures hotter than such radiation seen up close. In fact, the cosmic background radiation temperature measures hotter in direct proportion to the distance at which the radiation is measured.[49] (See figure 5.6, page 101.) Such an outcome establishes that the universe has been continuously cooling from a near infinitely hot origin. This result establishes that thousands of years ago the Bible correctly predicted every significant feature of current big bang cosmology.

UNIQUE VIEWING PLATFORM

Since the universe began about 13.73 billion years ago, cosmic expansion has increased the distance between star clusters, galaxies, and galaxy clusters by nearly a thousand times. But, because the speed of light has exceeded the speed of cosmic expansion (by an average of 100 times), the proportion of the universe visible to an Earth-bound spectator has increased at an even faster rate.

For the first time in cosmic history, light from the most distant galaxies and star clusters has reached the Milky Way Galaxy where human observers

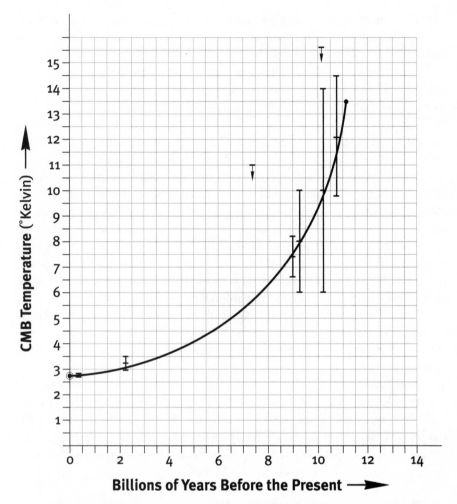

Figure 5.6: Evidence Supporting Hot Big Bang Creation Models
Today, the temperature of the radiation left over from the cosmic origin event measures 2.725° ± 0.001°C above absolute zero (absolute zero = -273.16°C). This graph shows temperature measurements from eight time periods in the history of the universe. Each measurement (represented by a small cross with its accompanying error bar) indicates the temperature of the cosmic background radiation detected in a distant gas cloud. The downward arrows show upper-limit measurements. Superimposed on this graph of actual measurements is the temperature curve predicted by hot big bang models.

can detect it. Even the radiation left over from the cosmic creation event (cosmic background radiation) is discernable. This is also the *last* era in cosmic history when such light and radiation will be visible.

Dark energy causes the universe's expansion to speed up as the cosmos ages.[50] This space energy density already has accelerated this expansion so much that objects formed just after the creation event are moving away from human observers at nearly the velocity of light. Soon dark energy will cause the expansion to accelerate beyond light's velocity. Then, distant objects will no longer be visible from Earth's vantage point.

As the universe expands at a progressively faster rate under the influence of dark energy, humans will see less and less of the universe. Humanity lives at the only moment in cosmic history when the first-formed stars, the galaxies, and even the cosmic background radiation can be observed. From an evolutionary perspective, such a circumstance is sheer coincidence. From a biblical perspective, the Creator timed humanity's moment in cosmic history to facilitate discovery of His existence and of His attributes through observation and consideration of the heavens.[51]

Humanity's arrival is not only perfectly timed but also perfectly placed. At virtually any other location in the Milky Way Galaxy, gas, dust, nearby stars, star clusters, and/or galactic arms would block a spectator's view. And, in almost every other medium-large spiral galaxy (the size required for life), nearby galaxies and/or galaxy clusters would obscure the view of a distant universe. Knowledge of the deep history of the universe would be hidden.

These findings of science align with the biblical framework of RTB's model. They correlate with God's purpose to make His existence and attributes visible to every human being through contemplation of the heavens. In Psalm 19:1-3, David wrote:

> The heavens declare the glory of God;
>   the skies proclaim the work of his hands.
> Day after day they pour forth speech;
>   night after night they display knowledge.
> There is no speech or language
>   where their voice is not heard.

SUPERNOVAE "COINCIDENCES"

Given the laws of physics and the gross characteristics of the universe (in particular its rate of expansion), formation of life-essential elements in adequate abundance takes a long time. The universe started with only one element, hydrogen. From three to three-and-a-half minutes after the beginning, nearly 25 percent of that hydrogen, by mass, converted to helium, plus trace amounts of lithium and deuterium (heavy hydrogen).[52] From four minutes onward, nuclear furnaces in the cores of large stars produced the rest of the elements.[53]

Rocky planets couldn't form and life chemistry wasn't possible until at least two generations of stars had developed, burned, and exploded their nuclear ashes into interstellar space. Even then, adequate abundances of elements required some highly specialized circumstances. First, production of the full range of life-essential elements in as short a time as 9.5 billion years (the time of Earth's formation) required the well-timed occurrence of three types of supernovae (relatively rare, massive stellar explosions) in proximity to one another: a type I, a normal type II, and an especially rare species of type II. Each produced a different suite of heavy elements.

Second, the three types of supernovae had to detonate not only close to one another but also near the gas and dust cloud where the sun had begun to coalesce. If one of them had exploded too close to this emerging solar nebula, that supernova would have blown the solar nebula apart, and no rocky planets capable of sustaining life could have formed. Any supernova that exploded too far away from the solar nebula would not have provided enough enrichment of certain heavy elements critical for advanced life chemistry.

Third, the timing of the three supernovae eruptions required precise orchestration. If any of them had exploded either too early or too late (relative to the emerging solar nebula) too few heavy elements from the three supernovae would have been incorporated into the solar nebula.

The level of fine-tuning necessary to explain the localized availability and adequate quantities of all advanced-life-essential heavy elements so early in cosmic history defies all probability. The odds of such a coincidence would be about the same as those of an explosion in a bicycle factory flinging out onto a sidewalk a block away, right in front of a cyclist, all the parts needed to assemble a new bicycle without doing any harm or damage to the parts, the cyclist, or the sidewalk. The intervention of a Creator intent on bringing humanity onto the cosmic scene at the earliest possible moment seems a more reasonable explanation.

ABUNDANT URANIUM AND THORIUM ON EARTH

Various heavy elements must be available in certain minimum abundances before life can exist. As the universe proceeds through successive generations of stars, the abundance of most heavy elements steadily increases—with one exception. The abundance of radiometric elements may increase or decrease depending on their decay rates.

Once the universe has aged by a few billion years, ongoing cosmic expansion begins to slow the rate of star formation. A slower rate of star formation means a slower rate of the kinds of stellar burnout events (primarily supernovae) that produce radiometric isotopes (nearly all atomic elements exist as a suite of isotopes—atoms with the same number of protons but with different numbers of neutrons). Thus, depending on the half-life of a particular radiometric isotope (that's the time required for exactly one-half of a certain amount of a radiometric substance to decay into its daughter products), the production of that isotope through stellar burnout events may eventually fail to keep pace with the decay of previously existing amounts of that isotope. Therefore, for each radiometric isotope, its abundance in the universe will steadily increase up to a peak value and thereafter steadily decline.

Several radioactive isotopes are essential to make Earth suitable for advanced life. Uranium-235, uranium-238, and thorium-232 play the most crucial roles (see chapter 7, pages 135–138 and pages 146–147). Based on measurements of the cosmic expansion rate, astronomers have determined that the cosmic abundance of these elements, collectively, peaked when the universe was two-thirds of its present age, some 4.57 billion years ago (see figure 5.7, page 105). This date precisely matches Earth's origin date (4.5662 ± 0.0001 billion years ago).[54]

Again, from a naturalistic perspective, the concurrence of Earth's origin with the moment when the radiometric isotopes most essential to advanced life reached their peak abundance seems bafflingly coincidental. From the perspective of a purposeful Creator, however, the precise timing of Earth's formation at the moment these radiometric isotopes reached their advanced-life-essential collective peak abundance would be expected. The discoveries again coincide with the biblical account.

SCIENTIFIC EVIDENCE PAINTS A PICTURE

Creation—truth or fiction? The cosmic beginning points toward a causal Agent, and the One behind the Bible's creation story fits the description well. The features of the heavens seem to speak of His existence and purpose.

The expanding universe fits the picture painted by the prophets Job, Isaiah, Jeremiah, and Zechariah, as well as by the psalmist. The great expanse of space and time was definitely not a waste—it permitted life-essential elements to come together so that Earth might form and be filled

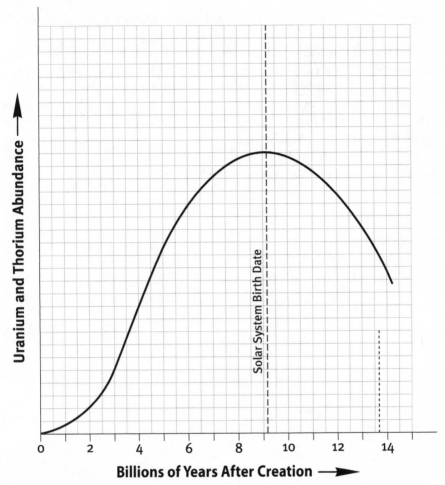

Figure 5.7: Uranium and Thorium Abundances in the Milky Way Galaxy
As the Milky Way Galaxy ages, the rate at which it produces uranium and thorium peaks and then declines according to the rise and decline of the supernova eruption rate. However, because uranium and thorium undergo radiometric decay, the production of additional quantities of these elements in supernovae eventually fails to compensate for the loss of uranium and thorium through decay. What's significant for Earth and for life is that the cosmic moment of their peak abundance coincides with the birth of the solar system.

with creatures both primitive and advanced. Hundreds of finely tuned characteristics in the cosmos made human life possible, suggesting the perfect plan and design of a transcendent God.

How can this scenario be explained? So, which is it? Given how accurately the Bible's cosmological content corresponds to cosmic reality, how precisely it foreshadowed the cosmic data, its statements with reference to other scientific disciplines warrant careful examination. The next chapter continues checking out the evidence by examining how life got its start. It uses the latest scientific findings about the origin and history of Earth to test the viability of RTB's creation model.

CHAPTER 6

PLANETARY SCIENCE AND LIFE'S ORIGIN TEST THE RTB CREATION MODEL

People today are as fascinated with the idea of life from Mars as they were in 1898 when they first read *The War of the Worlds* and in 1938 when they first heard it enacted on radio. With a price tag of $500 million, the Mars Reconnaissance Orbiter was launched August 12, 2005, to probe the mysteries of the red planet. This spacecraft is expected to return more than 40 times more data than any previous Mars mission. Anticipation of breakthroughs runs high, specifically with respect to life. As recently announced:

> Each of the scientific experiments will build on information from earlier missions to Mars and pave the way for future missions. They're all part of NASA's strategy to "follow the water" to look for signs of present or past life on Mars.[1]

Scientists know that any planet capable of sustaining life must have a stable supply of liquid water. Many other conditions also determine the suitability of a potential home for living creatures. Life's requirement and its fragility mean not just any place will do.

Astronomers generally concede that an extraordinary planet is needed for the possible support of life. However, given the vastness of the cosmos and the probable existence of billions and billions of planets, Carl Sagan's assumption that habitable planets—and life upon them—could be common may seem reasonable. But is it? What does the available evidence say about how life originates and about the abundance of planets suitable to sustain life?

The RTB creation model posits that Earth's first life arose and ultimately persisted on the planet not as a random, naturalistic outcome but rather as the product of supernatural intervention by the Creator of the universe. An outline of a few of the key elements of RTB's creation model as it pertains to life's origin and unique home appears in "Earth's Habitability and Origin-of-Life Highlights," below. Two previously published books, *The Creator and the Cosmos* and *Origins of Life*, explain in some detail this portion of the creation model with more to be presented in at least two future books.[2] For now, a brief overview of the model's biblical features follows, along with specific scientific assessments.

Earth's Habitability and Origin-of-Life Highlights

The RTB creation model for life's beginnings and unique habitat rests on the following biblical assertions and inferences:

- God personally shaped and crafted the universe to produce all the essential ingredients for life, ultimately human life. The heavens contain abundant evidence of God's fine-tuning for the sake of all life. Human observers thus have the potential to uncover evidence of the supernatural, super-intelligent design of life's cosmic environment for the specific benefit of humanity.
- God personally shaped and crafted Earth so that the planet would provide for humanity's physical needs and well-being. Earth contains abundant evidence of God's fine-tuning for the sake of human life and civilization. Human investigators thus have the potential to uncover evidence of the supernatural, super-intelligent design of Earth for the specific benefit and bountiful provision of the human species.
- God is the author of all life. God supernaturally intervened to create life early in Earth's history under relatively hostile conditions and to sustain that life in the optimal time, place, and form for the future benefit of humanity.

ARE WE ALONE?

Naturalists' optimistic projections of potentially abundant possible intelligent-life sites are rooted in an equation that dates back to 1960. Astronomer Frank Drake proposed that the number of intelligent-life-carrying planets in the Milky Way Galaxy alone would be the number of stars in the galaxy multiplied by the fraction of those stars that possess planets, multiplied by the average number of planets orbiting the planet-possessing stars, multiplied by the fraction of those planets sufficiently similar (for life-support) to Earth, multiplied by the fraction of life-supportable planets that actually contain life, multiplied by the fraction of life-containing planets that eventually produce intelligent life, multiplied by the fraction of intelligent-life-producing planets that currently possess intelligent life (see figure 6.1, below).

Figure 6.1: Frank Drake and His Famous Equation
In 1960 astronomer Frank Drake developed an equation to predict how many planets in the universe might be home to intelligent life. In that era, astronomers could only guess at the probabilities for the various factors in the equation. Today, however, scientists have been able to make observations and measurements revealing that the probability for some of these factors is indistinguishable from zero. Thus, the solution to the equation becomes indistinguishable from zero.

(Photo courtesy of Dr. Seth Shostak/Photo Researchers, Inc.)

In the 1960s only the first number in the Drake Equation had been determined. Drake, and later Carl Sagan, dealt with the lack of additional data by presuming either that the unknown fractions were equal to one-tenth or that the unknown numbers and fractions in the equation were the same as they were in Earth's planetary system. Therefore, according to Drake and Sagan, the number of stars in the Milky Way Galaxy with planets carrying intelligent life would be 300 billion times one-tenth times 10 times one-tenth times one-tenth times one-tenth times one-tenth, which equals 30 million planets.

Today, thanks to advances in planetary science and other disciplines, scientists no longer need to guess what numbers should go into the Drake Equation. New probabilities based on real observations, experiments, and laboratory and computer simulations yield more realistic numbers. And these results provide new means for testing the RTB creation model. To begin, the existence of planets outside the solar system is no longer pure speculation.

Extrasolar Planets

The first planet discovered in orbit around a stable burning star other than the sun was found in 1995.[3] Initially, expectations ran high that similar planet-bearing stars would prove common. Scientists also anticipated that planetary systems with life-support capacity would be similarly plentiful. Those hopes have since been dashed.

Today, more than 190 extrasolar planets are known to orbit stable burning stars.[4]

Most of these extrasolar planets are at least as massive as Jupiter (Jupiter = 318 Earth masses). Of course, very massive planets can be detected much more easily than small ones. However, these very massive planets with their ultra-thick atmospheres of inappropriate-for-life compositions are not candidates for the support of life.

And yet, a Jupiter-sized planet is essential for Earth's life to exist. In the solar system, Jupiter's position and enormous mass allows it to operate as a gravitational shield, protecting Earth from absorbing too many life-damaging collisions with asteroids and comets. If Jupiter were any less massive or any farther away, Earth would not receive the gravitational protection necessary to shield life over a long time period. On the other hand, if Jupiter were any more massive, any closer to Earth, had a less circular orbit, or came into any orbital resonances[5] with other gas giant planets, Earth's orbit would be so disturbed (by Jupiter's gravity) that it could not remain within a life-support zone (see "Climatic Runaways," page 112).

None of the extrasolar planetary systems discovered thus far to contain

Jupiter-sized planets manifests the just-right characteristics that would permit the existence of a life-support planet. Their Jupiter-sized planets are too massive, too close to their stars, or have orbits that are too eccentric (noncircular).

Astronomers have noted that nearly all of the planet-possessing stars discovered to date are very much like the sun. They are nearly the same mass and same composition. These characteristics are consistent with the best theoretical simulations for planet formation.

Scientists now know that a star must be characterized by a certain minimum abundance of metals for planet formation to be possible in its vicinity. (Astronomers refer to all elements heavier than helium as metals.) Only 2 percent of the Milky Way Galaxy's stars contain the necessary abundance of metals.[6] The Milky Way Galaxy itself is exceptional in that it contains a high population of metal-rich stars. Most galaxies contain only a miniscule number of metal-rich stars or none at all.

Too high an abundance of metals represents as much of a problem as too low an abundance. A too-high metal abundance (which many Milky Way Galaxy stars manifest) produces so many asteroids and tiny planets (planetesimals) that any Earth-sized planet that could possibly support life would suffer ongoing catastrophic-to-life collisions.

Far from proving the solar system typical or randomly structured, the discovery of extrasolar planets and the theoretical simulations prompted by these discoveries increasingly suggest that the solar system is exquisitely designed for life. The RTB creation model anticipates that as astronomers discover more extrasolar planets and learn more about the physics of their planetary systems, evidence will increasingly show that Earth's solar system features are rare and that many of its distinctives are amazingly fine-tuned for the needs of advanced life.

THE PERFECT COLLISION

Most of the time, a major crash constitutes a catastrophe. But one gigantic collision proved quite the opposite. When primordial Earth was only 30 to 50 million years old, a planet between 11 and 14 percent the mass of Earth (for comparison's sake, Mars = 0.107 Earth masses) smashed into Earth at a 45-degree angle. The speed of the planet upon impact was surprisingly low (less than 4 kilometers per second).[7]

This astonishing crash:

- Replaced Earth's thick, life-suffocating atmosphere with one thin enough (about 100 times thinner) for light to pass through. The

Climatic Runaways

Earth's biosphere remains poised between a runaway freeze and a runaway evaporation. If the mean temperature of Earth's surface dropped even a few degrees, more snow and ice than normal would form. Snow and ice reflect solar energy much more efficiently than do other surface materials. This reflection further lowers surface temperatures, forming more snow and ice. Subsequent temperatures would drop lower still, further exacerbating the problem.

If the mean temperature of Earth's surface warmed by just a few degrees, more water vapor and carbon dioxide would collect in the atmosphere. The extra water vapor and carbon dioxide would then create a more effective greenhouse condition. Earth's atmosphere would thus trap more heat, sending the surface temperature higher still, releasing more water vapor and carbon dioxide into the atmosphere. Surface temperatures would continue to cycle higher and higher.

The sensitivity of a planet's orbital distance and orbital stability demonstrates the astronomical fine-tuning required for life's sustenance. A change or disturbance of only 1 percent in Earth's orbital distance from the sun would generate either a runaway freeze or runaway evaporation, making life's survival impossible.

---

new atmosphere had the just-right chemical composition to foster life's existence.

- Boosted the mass and density of Earth just enough to allow Earth's gravity to retain a large, but not too-large, quantity of water vapor for billions of years.

- Raised the amount of iron in Earth's core closer to the level needed to provide Earth with a strong and enduring magnetic field (the remainder of the needed iron came from a later collision event—see pages 115–117). This magnetic field shields life from deadly cosmic rays and solar X-rays.

- Delivered to Earth's core and mantle quantities of iron and other associated elements in the just-right ratios to produce sufficiently long-lasting, continent-building plate tectonics. (Earth began without any continents.) Plate tectonics also perform a crucial role in compensating for the sun's increasing brightness (see chapter 7, pages 134–138) so that life can be sustained on Earth for billions of years.

- Increased the iron content of Earth's crust in a way that permits a huge abundance of ocean life that in turn supports advanced land life.[8]

- Played a major role in salting Earth's interior with an abundance of long-lasting radioisotopes, the heat from which drives most of Earth's tectonic activity and volcanism.[9]
- Gradually slowed Earth's rotation rate so that a wide variety of lower life-forms could survive long enough to provide the resources needed for the existence of advanced life-forms. This slowing was also directly significant to advanced life, which cannot survive the high wind velocities of a more rapid rotation rate.
- Stabilized the tilt of Earth's rotation axis, protecting the planet from rapid and extreme climatic variations.[10]

Each of these characteristics altered Earth to make conditions far more favorable for living creatures. Some of the changes, particularly the latter two, came about because of the most impressive impact result—*the creation of Earth's moon* (see figure 6.2, page 114).

Without the moon, Earth could not sustain advanced life. Only a massive, solo moon has the gravitational strength to stabilize the tilt of a planet's rotation axis for an extended period. However, with a mass only 2 percent larger (a radius just seven miles longer), the moon would have pulled the tilt of Earth's rotation axis out of stability. And the gravity of a moon with only 2 percent less mass would be too weak to put adequate brakes on Earth's rotation rate.[11]

Without that slowing effect, Earth would complete each spin—each day—in the blinding speed of just a few hours, instead of 24. Even a slightly faster rotation rate would generate overwhelmingly frequent, long-lasting, and powerful hurricanes and tornadoes. A slightly slower spin rate would bring life-threatening temperature swings from day to night and would drastically decrease rainfall on continental landmasses.

While it is true that a slightly less massive moon could still slow Earth's rotation rate to 24 hours per day, it would take considerably longer than about 4.5 billion years for that slowing to occur. The problem with such a scenario is that a technologically advanced species requires a sun to be in its most stable burning phase, which begins when the sun is about 4.5 billion years old.[12]

The collision that led to the formation of a just-right moon appears to have been perfectly aimed, weighed, and timed to change Earth from a "formless and empty" wasteland into a site where advanced life could survive and thrive. The degree of "anthropic" fine-tuning manifested in just this one event argues powerfully for a divine Creator.

Even if the observable universe contained as many planets as stars (10 billion trillion, or 10^{22}), probability theory would identify even one such

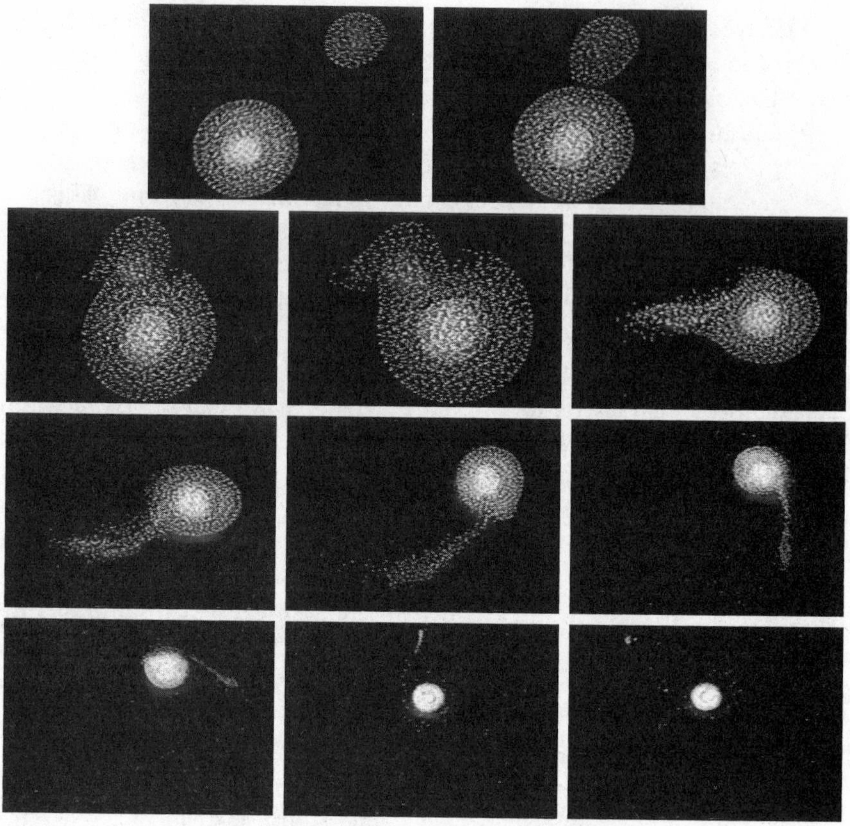

Figure 6.2: Formation of Earth's Moon
The frames depict the sequence of events in which the moon formed. They show
the before, during, and after of a Mars-sized planet's collision with Earth.

(Photo courtesy of A. G. W. Cameron, PhD, Donald H. Menzel
Research Professor Emeritus of Astrophysics, Harvard University College of Astronomy)

moon-forming collision as either impossible or miraculous. This crash
resulted in a planet with its surface gravity, surface temperature, atmospheric
composition, atmospheric pressure, iron abundance, tectonics, volcanism,
rotation rate, rate of decline in rotation rate, and stable rotation axis tilt *all*
in the just-right range to support advanced life.[13] Such an event seems far
more in line with the purposeful intent of a divine Creator, as set forth in
RTB's creation model, than with the claim that the formation of the moon
and the coincident transformation of the Earth were mere accidents.

This last point can be framed as a predictive test. Naturalistic models for
the universe and life's history would predict that as astronomers discover
extrasolar bodies as small as the Earth and moon and learn more about the

physics of planet and moon formation, relatively simple naturalistic explanations for the Earth-Moon system and its capacity to support advanced life will be found. The RTB creation model, on the other hand, predicts the opposite: that increasing knowledge will show that strict analogs to the Earth-Moon system demand even more fine-tuning than is evident today and that such analogs in other planetary systems will prove extremely rare or will not be observed.

A (LATE) FAVORABLE BOMBARDMENT

The collision that shaped the Earth-Moon system left Earth slightly short of the iron needed to maintain a strong magnetic field and to sustain adequate plate tectonics over the next 4 or 5 billion years. Does that shortfall undermine the case for a Creator or perhaps reflect an inability to get the conditions right? Science shows what happened next.

Between 3.9 and 3.8 billion years ago, a gravitational disturbance[14] in the solar system caused a shower of giant asteroids and comets to pelt the inner solar system.[15] (See figure 6.3, page 116.) The craters and "seas" on the moon, Mercury, and Mars (specifically, the numbers and sizes of craters and seas and their measurable erosion) testify to this event, as does the swath of comets and asteroids (the Kuiper Belt) lying beyond the orbit of Neptune. Because Earth is the most massive of the inner solar system bodies and therefore manifests the greatest gravitational attraction, it suffered the greatest impact damage. Astronomers estimate that roughly 17,000 collisions scattered a total of 200 tons of extraterrestrial material per square yard over the entire surface of Earth during this Late Heavy Bombardment.[16]

Recently, geophysicists in Germany reproduced in laboratory experiments the temperature and pressure conditions of Earth's mantle and core.[17] Their results demonstrated that this bombardment heated Earth's surface to such a degree that an ocean of hot magma several hundred kilometers deep covered the entire planet. In addition, their calculations and experiments showed that the rare large size and high oxygen content of Earth's core—critical factors for the long-term maintenance of Earth's magnetic and tectonic characteristics essential for advanced life—resulted from this deep magma ocean.

Just like the earlier collision event, the Late Heavy Bombardment required precise fine-tuning. One blast, by itself, was insufficient to prepare Earth for sustenance of advanced life. Both events had to be carefully fine-tuned and timed to ensure that Earth formed a core with the just-right chemical composition and size to maintain the magnetic field and plate tectonics required by advanced life.

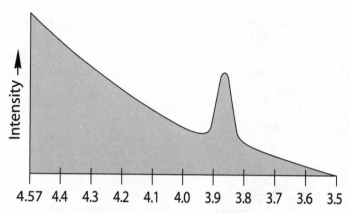

4.57 4.4 4.3 4.2 4.1 4.0 3.9 3.8 3.7 3.6 3.5

Figure 6.3: Early Bombardment of Earth
Large asteroids and comets pummeled the entire inner solar system (from Mars to Mercury) between 4.57 and 3.5 billion years ago. The intensity of the bombardment declined exponentially through that era — except for a brief episode known as the Late Heavy Bombardment (3.9 to 3.8 billion years ago). This late bombardment was triggered by an orbital resonance between Jupiter and Saturn that destabilized the Kuiper Belt's asteroids and comets.

(Presented by Marc van Zuilen in his lecture on July 1, 2002, at the International Society for the Study of the Origin of Life Conference in Oaxaca, Mexico, abstract #16)

The timing of the Late Heavy Bombardment compels researchers to conclude that life's origin on Earth occurred within a geological instant. Carbon-13 to carbon-12 ratio analysis of ancient carbonaceous material, along with abundance analysis of certain uranium oxide precipitates, establishes that life was abundant on Earth as far back as 3.8 billion years ago.[18] Carbon-13 to carbon-12 ratio analysis, plus nitrogen-15 to nitrogen-14 ratio analysis, on ancient carbonaceous material further establishes that a primordial soup or mineral substrate of prebiotic molecules never existed on Earth.[19] However, the Late Heavy Bombardment did not end until about 3.8 billion years ago. Therefore, Earth transitioned from life-exterminating conditions to a life-abundant state in less than a few million years.

Nontheistic models must attribute the timing and the phenomenal fine-tuning of both the moon-forming collision event and the Late Heavy Bombardment to extremely lucky coincidences. They also must assign the brevity and timing of life's origin *without benefit of any prebiotic compounds* to an unfathomably rare chance outcome (for calculations of just how rare a chance outcome, see the RTB book *Origins of Life*[20]). The RTB creation model, on the other hand, explains the two collision events as the handiwork of God, who chose, in the context of His physical laws, to prepare Earth as quickly as possible for the kinds of simple life-forms that in turn established an environment suitable for advanced life. The geologically instantaneous nature of life's origin—independent of prebiotics—and the abundance and diversity of life 3.8 billion years ago fit well with a Creator's plan to bring human beings upon Earth as quickly as possible and with all the resources needed to launch and sustain civilization and technology.

THE PARADOX OF LIFE'S BUILDING BLOCKS

From a naturalistic perspective, the origin of life required (among other things) a site where amino acids (the building blocks of proteins) and nucleotides (the building blocks of DNA and RNA) could be efficiently assembled and concentrated. Both oxygen and ultraviolet radiation are toxic to prebiotic chemistry. They powerfully shut down any possible synthesis of amino acids and nucleotides.

Because Earth's possible environments for life always had either oxygen or ultraviolet radiation, this shutdown effect explains from a naturalistic perspective why Earth contains no record of the existence of any prebiotics. (The presence of oxygen excludes prebiotic chemistry, whereas the lack of oxygen means no ozone shield can form in Earth's atmosphere to prevent the penetration of ultraviolet radiation from the sun.)

A few astronomers have speculated that perhaps some regions within dense interstellar molecular clouds might lack both oxygen and ultraviolet radiation and, therefore, could conceivably permit the formation of prebiotics. However, no extraterrestrial amino acids, nucleotides, or even the pentose sugars and nitrogenous bases that make up part of the nucleotides have been found there. Repeated claims for the detection of the simplest amino acid, glycine,[21] have proved false, as have claims for the detection of the simplest nitrogenous base, pyrimidine.[22]

A few meteorites contain up to six of the 20 biologically required amino acids at the level of several parts per million or less. One meteorite, the Murchison, contains three nitrogenous bases (at a few parts per billion). But research shows that the discovered amino acids and nitrogenous bases are not indigenous to the meteorites; rather, they result from terrestrial contamination—contact with the remains of life on Earth.[23] Furthermore, astronomers have found no amino acids or nitrogenous bases in comets, the source of the meteorites in question.[24] The absence of these prebiotics on Earth or anywhere in the Milky Way Galaxy[25] seems to rule out the possibility of life's emergence. And yet, life appeared—suddenly and early in Earth's history.[26] While this scenario has no reasonable explanation in naturalistic models, it is both anticipated and explained in RTB's creation model.

ON ONE HAND OR THE OTHER

Naturalistic models for life's origin also require a site where amino acids and pentose sugars of a specific orientation can be selected from the normal random mix of left- and right-handed configurations (see figure 6.4, below). Amino acids can link together to form protein chains only if the group is entirely one-handed (homochiral)—either all right or all left. Likewise, nucleotides can link together to form DNA or RNA only if all the pentose sugars are either all left-handed or all right-handed.

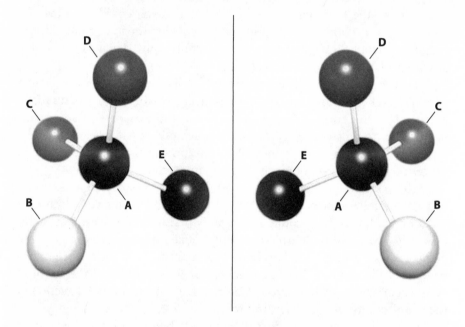

Figure 6.4: Mirror-Image Molecules and the Homochirality Problem
The building blocks of life molecules (amino acids and nucleotides) take either of two mirror-image configurations. This illustration shows the two possible configurations of a particular amino acid — one right-handed, the other left-handed. Life molecules can function, however, only if all the building blocks of a particular type are exclusively right-handed or exclusively left-handed, that is, "homochiral."

(Illustration courtesy of Digital Vision)

No known natural mechanism exists (present or past) on Earth for generating this homochirality.[27] While some individual mineral crystals can produce limited right-handed or left-handed enrichment (about 10 percent), any natural ensemble of mineral crystals contains roughly the same number of crystals that favor the production of left-handed configurations as it does the number that favor the production of right-handed configurations.[28]

Homochiral amino acids are difficult to produce even under highly complex and carefully controlled laboratory conditions. The only promising result has arisen from experiments exploiting 100 percent circularly polarized ultraviolet light. In one case, careful control of the intensity and wavelength of such light yielded as much as a 20 percent excess of left-handed amino acids.[29] The cost, however, of producing this excess was the destruction of nearly all of the original amino acids. In fact, for every 2 percent excess generated, half or more of the amino acids in the original mixture were destroyed.[30]

The only known natural source of circularly polarized ultraviolet (UV) light lies far from Earth—in the synchrotron radiation emitted by some neutron stars and also in the vicinity of some black holes. At best, such light is only slightly more than 40 percent circularly polarized and only for brief periods.[31] Experiments simulating the synchrotron radiation emitted by neutron stars produced a mere 1.12 percent excess of homochiral amino acids.[32] These experiments, however, used 100 percent circularly polarized light of one wavelength only. Neutron stars and the regions around black holes emit light across a broad spectrum, bringing the Kuhn-Condon rule into play. This rule notes that while one wavelength of circularly polarized UV light destroys more left-handed amino acids than right-handed ones, a different wavelength will have the opposite effect.[33] Thus, a broad band of light will fail to produce any significant homochirality.

The lack of any discernible natural mechanism for generating homochiral amino acids and pentose sugars remains devastating for naturalistic models for life's origin. Theistic evolutionists who hold that God's creative activity manifests itself only and always through natural processes face the same intractable problem. RTB's biblical creation model proposes that a purposeful Creator intervened to produce the otherwise impossible homochirality essential for construction of life molecules.

LIFE FROM BEYOND — OR NOT

The daunting problems facing a naturalistic scenario for the origin of life on Earth have given birth to a new research discipline—astrobiology. As the name implies, astrobiology is the scientific discipline concerned with life beyond Earth. It remains, as yet, a data-free discipline. No such life has yet been found or has even been shown possible.

Today, as this area of research grows at a stupendous rate in terms of funding, it has ruled out as potential life sites:

- Other solar system bodies
- Interplanetary gas and dust
- Interplanetary rocks and comets
- Interstellar gas and dust
- Interstellar rocks

In none of these places could life have originated on its own nor could any of these things have served as vehicles by which life on some distant planet or moon could have randomly hitched a ride to Earth.[34]

The lack of any naturally occurring life-transport mechanism to carry life to Earth from some hypothesized site in the Milky Way Galaxy has led some prominent origin-of-life researchers to suggest that life arose on Earth through "directed panspermia." According to this hypothesis, intelligent aliens intentionally seeded Earth with life, which they carried to this planet on spacecraft about 3.8 billion years ago.[35]

The directed panspermia proposal collides head-on with three insurmountable barriers. First, the laws of physics and the space-time dimensions of the universe would prevent physical aliens, no matter how well funded or technologically advanced, from safely traversing the necessary interstellar distances in any reasonable time period. Either their spacecraft or the life on board (or both) would be destroyed in transit by radiation, interstellar dust and debris, and long-term wear and stress on the spacecraft and its occupants.[36]

Second, the universe at the time life appeared on Earth was only 9.9 billion years old—far too young for an advanced physical species to have emerged and developed technological sophistication by any conceivable natural means. The building blocks of advanced life and the resources to sustain advanced civilization simply didn't exist early enough in the cosmic maturation process.

The third and most glaring weakness of directed panspermia is that it doesn't solve the origin-of-life problem. Directed panspermia merely transfers the problem to a different time and place.

The RTB creation model, however, confronts all three obstacles head-on. Directed panspermia may be one useful way to describe the work of an intelligent, transcendent "alien"—the biblical Creator. His existence is unconfined by the physics, dimensions, funding, technology, or age of the universe. The God of the Bible simultaneously transcends the universe He created[37] and has access to every space-time locale within it by His immanence.[38] The Creator could have brought together life's essential ingredients from materials in and around Earth, then kindled the spark of life in them by the same power He used to ignite the big bang.

A number of scientists have proposed an alternative to acknowledging the God of the Bible as the answer to the origin-of-life question. They hypothesize the existence of a yet undiscovered, self-organizing, complexity-enhancing law of physics as the "cause" of life.[39] Such a law, however, would directly contradict the second law of thermodynamics, which says that all systems in the universe proceed toward increasing disorder and decreasing complexity (see page 73). Although isolated departures from thermodynamic equilibrium can occur, the more extreme the departure, the more rapid the return to equilibrium. For anything as complex as a simple bacterium, the return-time would be so fast as to be indistinguishable from zero.

Evidence in support of the anthropic principle actually rules out any possibility of a self-organizing, complexity-enhancing law of physics as the cause of life. For stars and planets to form and, thus, for any kind of physical life to be possible anywhere and at any time in the universe, the universe must be both homogeneous and uniform to a very high degree. All parts of the universe must also be subject to a very high measure of entropy (increasing disorder).

Astrobiology, with its astronomically large and growing support from both public and private sectors, has yet to produce a single piece of evidence for indigenous extraterrestrial life. Eventually, however, researchers will find life—the remnants of Earth life—on other solar system bodies. During the past 3.8 billion years, impacts by large meteorites have scattered hundreds of millions of tons of Earth material throughout the solar system.[40] This dust and debris contains thousands of tons of organisms and their remains, and much of this material would have landed on the nearby planets and their moons (see table 6.1, page 122).

Two places where some of Earth's microorganisms may have been temporarily viable are Mars and Earth's moon. Because of their proximity to Earth, a remote possibility exists that Earth life could have arrived on the moon or Mars aboard a speck of meteoritic material before the harsh lunar or Martian conditions killed it.

EVEN EARLIER LIFE?

The title of the RTB book *Origins* [plural] *of Life*[41] was chosen to allow for the possibility that life originated by the Creator's hand more than once, before and/or after the Late Heavy Bombardment. Figure 6.3 (page 116) illustrates bombardment by large asteroids and comets that would have kept conditions unstable for life during the first billion years of Earth's history.

TABLE 6.1:
DELIVERY OF EARTH MATERIAL TO OTHER SOLAR SYSTEM BODIES

Meteorites massive enough to generate craters 100 kilometers or more across will cause Earth rocks and dust to escape Earth's gravity. Over the course of the solar system's 4.566-billion-year history, significant amounts of Earth material have likely accumulated on various solar system bodies. The amounts have been calculated as follows:[42]

SOLAR SYSTEM BODY	AMOUNT OF EARTH MATERIAL (GRAMS PER SQ. KILOMETER)
Moon	200,000 grams
Venus	3.1 grams
Mercury	3.3 grams
Mars	1.0 grams
Jupiter	0.09 grams
Saturn	0.01 grams

Clearly the moon-forming collision that occurred about 4.50 billion years ago (see pages 111–115) and the Late Heavy Bombardment about 3.85 billion years ago (see pages 115–117) made the entire planet temporarily uninhabitable for any conceivable life-form. While other total extermination events likely occurred at several epochs between these two major events and possibly a few more between 3.8 and 3.5 billion years ago, relatively life-benign conditions may have existed for brief interludes between them.

Indeed, scientists have recovered a few tiny zircon crystals dating to various epochs between 4.4 and 3.9 billion years ago with oxygen-18 to oxygen-16 ratios indicating that Earth possessed for at least a few brief time spans a watery ocean.[43] Thus, the possibility remains that life could have existed during these episodes.

Multiple origins of life during the few brief benign epochs between 4.4 and 3.9 billion years ago might have served one or more long-range purposes:

- Carbon dioxide, water vapor, and/or methane outgassing from Earth's interior and/or the input of such gases from comet collisions might have threatened such a buildup of greenhouse gases that Earth could have become permanently too hot for life. Early life-forms that efficiently consumed these greenhouse gases could have helped prevent such a catastrophe.

- Earth's early oxygen sinks (oxygen-consuming mineral reservoirs) may have been larger than current estimates suggest. In that case, early photosynthetic life would have helped fill the oxygen sinks so that humans could arrive at the best possible time (that is, early enough) for survival and for the launch of advanced civilization.
- Oxygenic photosynthetic life (photosynthetic life that requires at least some oxygen in the environment) dating back to the period just after the end of the Late Heavy Bombardment may have needed previous generations of anoxygenic photosynthetic life (photosynthetic life that does not require the existence of oxygen in its environment) to boost atmospheric and oceanic oxygen levels.
- The young sun's loss of mass may have proceeded at such a low rate that the buildup of greenhouse gases on Earth from outgassing and cometary input would have produced a potential imbalance. That is, the rate of cooling from solar mass loss may not have matched the rate of warming on Earth's surface from the buildup of greenhouse gases. Here, the just-right life-forms in just-right quantities at the just-right times could have corrected the imbalance.

The discovery of multiple origins of life on Earth, if such a discovery were made, would pose no problem for the RTB creation model. Such a finding could be shown to fit the purposes and powers of the biblical Creator. By contrast, such a discovery would only exacerbate the already intractable problems naturalistic models must face. Instead of needing to explain just one "virtually impossible" origin of life, they would need to explain several.

An Impossible Planet Made Possible

Without a preconceived commitment to naturalism, NASA's attempt to "follow the water" to extraterrestrial life sites seems a less than prudent investment. A good financial consultant would advise investors to put funding toward pursuits more likely to yield results such as research into these projects: Earth's unique timing and positioning in the universe, Earth's moon and its significant role in sustaining life, as well as the extraterrestrial bombardments that prepared Earth for life. Other valuable research could explore the presence of oxygen and ultraviolet radiation in and on early Earth, the possibility of some as yet unknown natural source for important prebiotics, and the homochirality challenge, among many others. So far, all the data point toward only one planet—Earth—prepared and fashioned for life.

RTB's creation model acknowledges the sequence of probability-defying events as phenomena guided and timed by the God of the Bible. The Creator prepared Earth efficiently and effectively for the kinds of simple life-forms that in turn established an environment fit for advanced life and ultimately for human civilization.

Either an impossibly lucky convergence of coincidences or something beyond the universe must account not only for such a scenario but also for life's beginning. The next chapter examines how evidence for the history of life on Earth compares with RTB's creation model.

CHAPTER 7

LIFE'S HISTORY ON EARTH TESTS THE RTB CREATION MODEL

D etails about the time, place, and crash scene added to *The War of the Worlds* broadcast's realism and emotional grip, especially on listeners with connections to the cities and towns identified. But most people were soon able to shake off the spell of that sensationally vivid portrayal. Visions of the real world quickly replaced the fleeting images of an alien invasion.

When questions about life's history arise, many people recall their high school biology textbook and its portrayal of the fossil record—a tree-of-life diagram showing how life branched forth from a single, simple life-form over several billion years. Gradually, through natural mutations and natural selection, this single-celled organism, which purportedly arose on its own from nonlife, developed into progressively more advanced life-forms until finally the human race stepped onto the scene. Presumably that progression continues.

In light of this neat diagram, the biblical creation story, as popularly (mis)understood, may seem as fanciful as the account of a Martian space-craft landing. For some students, it's "end of story," the creation story, that is. For others, it's the tension of compartmentalizing religion and science. To still others comes an inkling that there may be more to both stories.

And so there is. The latest research yields new evidences that demand a radical rearrangement of the branching tree diagram. At the same time, new understanding of Earth's and life's history reflects intricate preparations for humanity's benefit. The body of data defies Richard Dawkins' meta-phor of a Blind Watchmaker who somehow stumbles to the top of Mount Improbable.[1]

The RTB creation model proposes that the Creator of the universe super-naturally intervened on more than a few occasions to transform the earthly environment into a habitat suitable for the support of progressively more advanced species of life. The model also asserts that the Creator personally intervened to create all the advanced life species that have existed on Earth. An outline of the key elements of RTB's creation model as it pertains to life's history and habitat may be found in "Life History Highlights" below. Some details of this part of RTB's creation model are presented in *The Genesis Question*,[2] with more to come in future publications. A brief overview of distinctive features of that model for life's history appears in the pages to follow, along with specific scientific evaluations and tests.

Life History Highlights

The RTB creation model, as it pertains to life's history on Earth and the pre-paration of its habitat, is based on the following biblically derived premises:

- God personally shaped and crafted the Earth, including its interior, exterior, and atmosphere, over the entire history of life on Earth so that all the needs of every life-form God created would be provided. In doing so, He shaped and crafted Earth and all its life to supply everything for the launch and maintenance of human civilization.
- The fine-tuning of the physical environment for life and the precisely timed layering of life reveals to human investigators abundant and increasing evidence of the Creator's involvement. Human researchers thus have the potential to uncover accumulating evidence for the supernatural, super-intelligent design of the planet and its creatures over the entire history of Earth for the specific benefit of the human race.
- God created life in a progression from relatively simple to radically advanced. Research would be expected to increasingly confirm this pattern in life's history.
- God created three distinctly different forms of life, in this order: first, purely physical life; second, life that is both physical and soulish (manifesting mind, will, and emotions); and finally, one species with body, soul, and spirit. As with the origin of first life, the appearance of the first soulish life and the first spirit life was sudden and miraculous.
- God created life with optimal designs. Research should reveal

(continued)

increasing evidence of optimal design at all levels: in life's molecules, cells, tissues, organs, appendages, organisms, species, and ecologies. Because what is optimal for one species may be optimal for others, investigators should find evidence of design repetition, or common design features.

- God consistently created life with optimized ecological relationships. All the life He created — bacteria, plants, herbivores, carnivores, and parasites — was designed to interrelate in a manner that enhances the quality of life for all.

- God, as the author of all life, holds authority over all life. He determines when various life-forms enter and leave Earth's scene. Researchers can expect to find evidence that the timing of speciation and extinction events in the fossil record follows a strategic plan, a plan that prepares for humanity's future needs and well-being — and arrival at the best possible time for the launch of global civilization.

FIRST LIFE: EARLY, DIVERSE, AND COMPLEX

All prominent naturalistic models for life's history on Earth rest on the proposition that life is descended from the last universal common ancestor (LUCA). This single life-form is presumed to be much simpler than any life that exists on Earth today. The RTB creation model anticipates a different scenario: given His stated plans and purposes for the human race, the Creator would more likely "jump start" early life toward achievement of His specific goals.

Origin-of-life research increasingly agrees with the latter story. Evidence now shows the simultaneous appearance of multiple, distinct, complex unicellular life-forms rather than just a single, ultra-simple organism. This ensemble includes, at minimum, both oxygenic and anoxygenic photosynthetic life, several species of sulfate-reducing bacteria, and a diversity of other chemoautotrophs (bacteria that exploit high-energy chemicals to sustain metabolic reactions). Each primordial life-form played a crucial role in preparing the way (with remarkable efficiency and speed) for the eventual appearance of more advanced animals, ultimately human beings, and finally for the rapid emergence of civilization.

Genomics research reveals that no organism, not even a simple parasite (dependent on other species for some of its life-critical functions), can survive without at least 250 functioning gene products.[3] It came as a shock, then, when geochemists found uranium oxide precipitates in rocks older than 3.7 billion years. This discovery revealed that oxygen-exploiting photosynthetic bacteria were already prolific at that early date.[4] Such bacteria require more than 2,000 gene products—500 more than the simplest independent (nonparasitic) organisms alive today.[5]

A naturalistic explanation for Earth's first life demands that such life be orders of magnitude simpler and less diverse than the simplest independent life-forms on Earth today (bacteria and archaea). However, the sudden simultaneous appearance of highly diverse and not-so-simple life-forms as early as 3.8 billion years ago would be consistent with the work of a supernatural, super-intelligent Being, aggressively preparing Earth for humanity and a technologically advanced human civilization.

SPECIALIZED BACTERIA TRANSFORM POISONS

On Earth today, the sulfate-reducing bacteria and photosynthetic bacteria that dominated the geologic record from 3.8 to 2.9 billion years ago still play a critical role in some life-sustaining processes, such as the sulfur and carbon cycles.[6] Early in life's history sulfate-reducing bacteria participated directly in reforming and redistributing heavy elements that otherwise would have thwarted the existence of advanced life.

The primordial salting of Earth with heavy elements produced globally distributed deposits of arsenic, boron, chlorine, chromium, cobalt, copper, fluorine, iodine, iron, manganese, molybdenum, nickel, phosphorous, potassium, selenium, sulfur, tin, vanadium, and zinc—all of which are among life's vital "poisons."[7] Advanced life requires that certain trace amounts of these elements be present in the environment in soluble forms. Too much of these elements, dispersed too widely, however, would prove deadly.

Certain sulfate-reducing bacteria remove from water its low but toxic concentrations of these elements. For example, some species of bacteria consume water-soluble zinc and from that zinc manufacture pure sphalerite (ZnS).[8] This sphalerite is insoluble and, therefore, nontoxic for advanced life. Moreover, when sufficiently large, dense populations of these bacteria die and settle onto ocean and lake bottoms, they precipitate highly economic ZnS ore deposits.

Researchers now recognize that sulfate-reducing bacteria supplied much, if not all, of the concentrated (thus economic to mine) ore deposits of iron, magnesium, zinc, and lead available to humans. Ores of trace metals such as silver, arsenic, selenium, and other vital poisons may similarly owe their concentrations (and accessibility) to sulfate-reducing bacteria.

These bacteria performed an invaluable function on behalf of later, more advanced life-forms: they reduced soluble concentrations of specific elements in Earth's environment from poisonous levels to advanced-life-sustaining levels. From 2.9 billion years ago to the present, the abundance and diversity of sulfate-reducing bacteria declined to the just-right levels to maintain a delicate balance. Just enough of each poisonous heavy element remains in soluble form to nourish advanced life, but not too much returns in soluble form through erosion processes to harm advanced life.

The dominance of sulfate-reducing bacteria for nearly a billion years or more early in life's history paved the way for advanced life—not in a random way, but rather in an apparently purposeful way as though anticipating humans. Concentrated ore deposits equip humanity for a high-tech civilization. Without easy-to-mine ores of zinc, tin, molybdenum, iron, silver, magnesium, lead, and so forth, humans would likely have remained limited to Stone Age technology.

CRUSTY COLONIES TRANSFORM LANDMASSES

Detailed analyses of cryptogamic crust material—soils comprised of photosynthetic or oxygen-producing bacteria, fungi, mosses, sand, and clay existing in symbiotic colonies—reveal that such microbial soils dramatically transformed both the temperature and the chemistry of Earth's early landmasses (see figure 7.1, page 130). This material prepared the way for more advanced vegetation.[9] These findings solve a long-pondered biological puzzle—the late emergence of advanced land vegetation (about a half billion years ago).

Earth's early landmasses were relatively hot and soil-deficient. Cryptogamic colonies, including those that exist today, can withstand these harsh conditions. They effectively limit erosion while, at the same time, they enhance chemical conditioning of the soil, cool things down, and oxygenate the atmosphere. These microbial colonies took hold very early, on the few pockets of loose rock that existed on the first barren continental masses (see pages 134–137).

Over 2 to 3 billion years, cryptogamic colonies transformed these pieces of land into the large accumulations of stable, nutrient-rich soil that vascular plants (higher plants with conducting tissue consisting primarily of xylem and phloem) require. This function helps explain the long wait—roughly 3.3 billion years—for the arrival of the first advanced lifeforms. Again, it suggests anticipation of later life's needs.

FAINT SUN PARADOX

Study of an intriguing paradox has shown astronomers how the temperature at Earth's surface has remained ideal for life over the past 3.8 billion

Figure 7.1: A Cryptogamic Crust
Cryptogamic, or cryptobiotic, crusts are organic soils composed of rock, dirt, cyano-bacteria, lichens, microfungi, and various algae and bacteria. Over a 2- to 3-billion-year period, these symbiotic colonies transformed the surface of early Earth's land masses into soils suitable for the support of advanced plant life.

years despite the fact that the sun's luminosity (or brightness) has varied by as much as 15 percent or more. These findings are crucial for consideration in creation/evolution debates. They directly challenge a foundational assumption in many evolutionary models—that the sun can be counted on to provide stable temperature conditions through billions of years. The need to compensate somehow for variations in the sun's luminosity might help explain why the fossil record looks the way it does. It also elucidates the question of how much foresight and planning went into the frequent but intermittent introductions of life on Earth.

Astronomers have learned that the sun shed between 4 and 7 percent of its primordial mass during its first billion to billion and a half years.[10] Because the sun's luminosity is proportional to slightly less than the fourth power of its mass, its brightness during that era declined by about 15 percent. Since that low point, however, the sun's luminosity has steadily increased (see figure 7.2, page 132). Its increase results from its nuclear burning of hydrogen into helium. The helium produced through nuclear burning increases the core density of the sun, which causes a higher core temperature that in turn augments the rate of nuclear burning. Over the last 3.5 to 3 billion years, the sun's luminosity has increased by about 15 percent.[11]

This initial drop in the sun's brightness followed by a gradual increase should have destroyed any chance for life's tenure on Earth. Life cannot tolerate even a 1 percent change in incident luminosity (see "Climatic Runaways," chapter 6, page 112).[12]

These dramatic changes in the sun's luminosity obviously did *not* sterilize Earth. The capacity for the Earth's atmosphere to trap heat (the greenhouse effect) has a modulating effect on the sun's luminosity as it impacts Earth's surface. A faint sun can be compensated for by a more efficient greenhouse effect in Earth's atmosphere, while a less efficient atmospheric greenhouse effect can compensate for a brighter sun.

During roughly the first half billion years of life's history on Earth, life progressed from unicellular organisms well suited for relatively high heat (50–80°C, 120–175°F) to those well suited for relative cold (0–20°C, 32–68°F). Though at the time of life's origin 3.8 billion years ago the sun would have been considerably fainter than it is today, Earth's atmosphere was much richer in carbon dioxide and water vapor. Both carbon dioxide and water vapor are effective greenhouse gases. Consequently, Earth's surface would have been considerably warmer than it is today.

Given the frequent mass extinction events (such as asteroid collisions, supernova eruptions, gamma-ray bursts, and solar flares) at the time of Earth's early life, species of life that were driven to extinction could

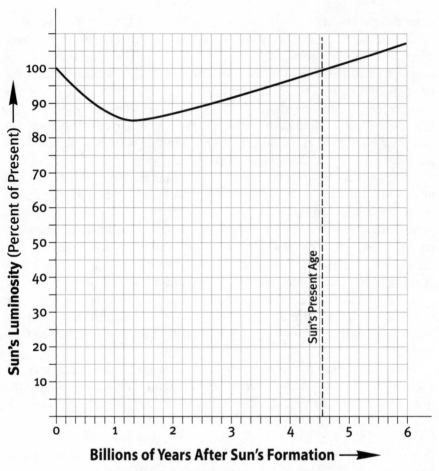

Figure 7.2: Luminosity History of the Sun
When the sun was young, it lost a small amount of its mass, enough to cause a (roughly) 15 percent loss of its light radiation. Thereafter, as its nuclear furnace converted more and more hydrogen into helium, the sun slowly recovered that lost luminosity. This brightening continues to this day and should continue for the next few billion years.

have been replaced with new species better suited to the cooler conditions brought on by the sun's declining luminosity. By itself, however, the replacement of heat-loving bacterial species with cold-loving bacterial species would not have been sufficient to accommodate (for long) the decreasing luminosity of the sun. A more important factor would have been outgassing from volcanic eruptions as they pumped just-right quantities of additional greenhouse gases into the atmosphere.

The number of just-right outcomes converging at the just-right times to compensate for the decreasing brightness of the youthful sun seriously strains naturalistic models. Unless the just-right life-forms appear at the just-right times, life is quickly extinguished. This problem can also occur if the just-right life-forms aren't removed at the just-right times. Finally, the amount and kind of volcanic outgassing must be just-right throughout the time the sun is losing mass. This much exactness seems inconsistent with the operation of mindless natural processes. It is perfectly consistent, however, with a super-intelligent, super-powerful Creator who knows the precise details of the sun's physics and the Earth's interior physics.

The sun's transition from dimming to brightening some 3.5 to 3 billion years ago necessitated a shift in direction for the type of life and for life's adaptation. A not-so-simple reversal back to heat-loving life-forms could have extended life's survivability on Earth by a few tens of millions of years at most, but this reversal also would have switched the direction of life's progression from more complex life to less complex life. A different strategy was needed, but one much more elaborate.

GREENHOUSE GAS REMOVAL

Maintenance of the necessary temperature conditions to sustain progressively more advanced life required the removal of greenhouse gases from Earth's atmosphere in direct proportion to the increase in the sun's luminosity. The erosion of silicates (the main components of continental landmasses) and the burial of organic carbon accomplished this intricate regulation.

In the erosion process, water from rain, streams, and mist catalyzes the chemical reaction between silicates and atmospheric carbon dioxide (a greenhouse gas). The end products are carbonates and silicon dioxide (sand). To get the necessary amount of exposed silicates, efficient plate tectonics must build up islands and continents (see pages 134–138). The rate of silicate erosion depends on these seven factors:

- Earth's rotation rate
- Average global rainfall
- Average global temperature
- Chemical composition of Earth's atmosphere
- Total area of Earth's landmasses
- Average slope of Earth's landmasses
- Quantities and types of plants on the land

Organisms, in particular photosynthetic plants, bacteria, and methanogens (methane-consuming bacteria), take carbon dioxide, water, and methane from the atmosphere and chemically transform them into sugars, starches, fats, proteins, and carbonates. If the sugars, starches, fats, proteins, and carbonates get buried (by erosion, tectonics, and/or volcanism) before they decay or get eaten by other organisms who then get buried before they decay, then greenhouse gases are converted into biodeposits through physical and chemical processes operating in Earth's crust.

The end products of greenhouse gas removal—coal, oil, natural gas, limestone, marble, gypsum, phosphates, and sand—are all valuable resources for launching and sustaining human civilization and technology. Without billions of years of greenhouse gas removal from Earth's atmosphere and the conversion of such gas into biodeposits, human civilization may never have taken off, nor would it have achieved its current level of technology.

Though exact proportions differed in the past, today about 80 percent of greenhouse gas removal takes place through silicate erosion and 20 percent through burial of organic material. Fine-tuning the removal of greenhouse gases to compensate for the increase in solar luminosity requires the fine-tuning of all seven factors that govern silicate erosion plus all the factors that govern the abundance, diversity, growth, decay, extinction, and burial of organisms. Furthermore, all this fine-tuning must be exquisitely timed throughout the past 3.5 to 3 billion years.

This continual fine-tuning over an extended time period challenges any reasonable naturalistic explanation. On the other hand, this fine-tuning suggests that the causal Agent behind Earth's life anticipated in intricate detail the future physical and chemical conditions of the sun and the Earth—not to mention the moon, the prime regulator of Earth's rotation rate. This exquisite fine-tuning fits well, too, with the biblically stated purposes for God's creating just-the-right life-forms at just-the-right time (and removing life-forms that are no longer appropriate) to sustain a habitable Earth and prepare the environment for the arrival of human beings. It provides insight to why the fossil record looks the way it does.

FORMATION OF CONTINENTS

The work of cryptogamic colonies in preparing Earth for advanced life (see page 129) is predicated on the existence of continental landmasses. Only with a combination of both large continents and large oceans can Earth's surface and atmosphere conditions possibly be regulated to compensate for the increasing luminosity of the sun. Such a combination is critical, too, for the existence of a high-tech species. However, Earth would have remained a permanent water world—water covering the entirety of its surface—if it were not for powerful and sustained plate tectonics.

Continents form out of light, silicate rocks that "float" higher above Earth's mantle than the denser basaltic rocks that comprise ocean floors. The separation of Earth's primordial crust into silicates and basalts occurs through the dynamics of crustal plate pressures and movements.

For plate tectonics to occur, Earth requires three things: (1) a powerful, long-lasting source of radioactive decay in its interior; (2) a stable, efficient dynamo (electric generator) at its core; and (3) an abundant supply of liquid water on its surface. The availability of each of these essentials in turn requires precise fine-tuning.

The presence of abundant, long-lived radioactive elements, for example, demands (among other things) a concert of highly unlikely, perfectly timed and placed supernova events as well as the miraculous collision and bombardment that formed the moon. The stability and efficiency of Earth's dynamo depends on the exact regulation of seven other major geophysical features:

- The relative abundances of silicon, iron, and sulfur in the solid inner core
- The viscosity at the boundaries between the solid inner core and the liquid outer core, and the liquid outer core and the mantle
- The ratio of the inner core radius to the outer core radius
- The ratio of the inner core magnetic diffusivity (a measure of how well a magnetic field diffuses throughout a conducting medium) to the outer core magnetic diffusivity
- The magnetic Reynold's number (a measure of viscous flow behavior) for the outer core
- The gravitational torques from the sun and moon
- Earth's core precession frequency[13]

The intricate process of continent building (one of the great wonders of the primordial era) began with the decay of radioactive elements in Earth's interior. This release of energy provided enough heat to generate convective cells throughout Earth's mantle. These cells are giant eddies that circulate in the mantle from just above Earth's core to just under Earth's crust. Different eddies associate with different regions of the crust. At the boundaries of these affected crust regions, subduction (the sliding of one crustal plate under another) can occur, but *only* if a huge amount of high-density, low-temperature liquid water is available to lubricate the sliding.

Subduction is governed by the rate at which minerals in the area where two underwater plates come together (the subduction zone) chemically react with water to form hydrated minerals.[14] The hydration process in the downward moving slabs leads to production of a talc layer that reduces and stabilizes the sliding friction between adjoining plates. A just-right level of lubrication permits efficient movement of one tectonic plate under another.

At these subduction zones, some basaltic slabs become hydrated. The newly hydrated minerals—silicates—are lighter than their nonhydrated cousins, and they have a lower melting point. Once these silicates began to float above the nonhydrated basalts, Earth's mountains and continents began to form. And, given their lower melting point, the silicates stayed liquid at depths closer to the surface, thereby forming volcanoes.

The gradual development of mountains and volcanoes eventually resulted in landmasses poking up above the surface of the water. With yet more time, these landmasses grew to become continents (see figure 7.3, page 137). However, they must continue growing at the just-right rates throughout most of life's history *if* the increasing luminosity of the sun is to be properly compensated for the sake of life.

For there to be any hope of removing enough greenhouse gases from Earth's atmosphere to compensate for the sun's increasing luminosity, the buildup of continental landmasses through plate tectonics must initially exceed and later at least keep up with the reduction of continental landmasses through erosion. However, the energy release from radioactive decay, the primary driving force behind plate tectonics and continental buildup, declines over time. The level of plate-tectonic movement today is only about a fifth of what it was when life first appeared on Earth.

However, the same collision that helped enrich Earth with radioactive elements also produced its gigantic moon. Coming around full circle in this amazingly complex scenario, the moon acts as a tidal brake on Earth, gradually slowing its rotation rate. A slower rotation rate means less erosion.

So many factors must be fined-tuned to sustain relatively large, stable silicate continents—not to mention Earth's large oceans—that defining their existence as a miracle seems fully justified. Even nontheistic scientists acknowledge Earth's long-lasting, large continents and oceans as truly amazing. Earth may indeed be the only planet in the universe with such features.[15]

More Oxygen to Breathe

The extremely high abundance of free oxygen in Earth's atmosphere dramatically sets Earth apart from all other heavenly bodies. Scientists

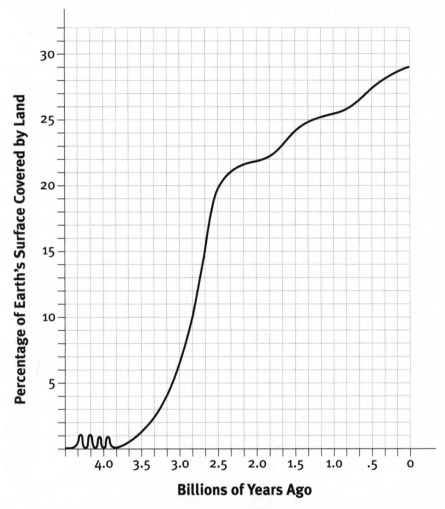

Figure 7.3: Growth of Continental Landmasses
For the first 750 million years of its history, Earth alternated between total and partial water coverage, with just a few small islands occasionally popping up above sea level. Later, as a result of ongoing plate-tectonic activity, continental landmass grew. Today, continental build-up through tectonic activity is only slightly greater than continental shrinkage due to erosion.

know only one way for a planet or a moon to accrue a large percentage of free oxygen in its atmosphere—through the long-term activity of super-abundant photosynthetic life. And yet even with the help of photosynthetic life, the job of oxygenating a planet's atmosphere to the level where it can support advanced life is not easy.

The role cryptogamic colonies played in oxygenating Earth's atmosphere, while significant, is minor compared to that of photosynthetic bacteria in the oceans. The abundance of these marine bacteria is so enormous that it could have transformed Earth's atmosphere from 1 or 2 percent oxygen to about 20 percent oxygen in only a few million years, perhaps less. At the same time, however, Earth's enormous oxygen sinks swallowed up so much oxygen as to slow that process by 3 billion years.

Erosion of the earliest rocks gradually delivered unoxidized iron and sulfur to the ocean where oxygen produced by photosynthesis reacted with them to form oxide deposits. Only after several global cycles of erosion, deposition, and tectonic uplift would iron and sulfur become fully oxidized.

Earth's mantle, the soft layer between Earth's crust and core, gobbles up still more oxygen. Only after several global cycles of volcanic lava deposition, then erosion, then tectonic subduction of crustal plates into the mantle, then more volcanic eruptions do the unoxidized mantle minerals become fully oxidized.

How long did it take to fill all of Earth's oxygen sinks? Geochemical analysis of preserved deep-water marine sediments indicates that the oceans became fully aerobic (oxygenated) sometime between 1.0 and 0.54 billion years ago.[16] The timing of Earth's oceanic and atmospheric oxygenation (see figure 7.4, page 139) helps explain the timing of the explosive appearance of complex life—the sudden, widespread, and extremely diverse existence of more than 40 phyla of complex animals.

No naturalistic model as yet can reasonably explain such an event, but this explosion fits well with several premises of RTB's creation model. The most important of these is that God initiated various life-forms over Earth's long history according to a well-thought-out plan that culminated in the creation of humanity.

BIOLOGY'S BIG BANG

Before 543 million years ago, Earth's zoo featured nothing more complex than cryptogamic colonies and some primitive sponges and jellyfish. Then, in a time window narrower than 2 to 3 million years (possibly much briefer), some 40 or more phyla of complex animals appeared, including 24 or 25 of the 30 animal phyla that remain on Earth today. (A phylum includes all animals with the same basic body pattern. For example, one phylum, the chordates, includes all animals with a central neural tube inside an outer tube. All vertebrates and some invertebrates are chordates.) Paleontologists call this sudden appearance of nearly all known types of

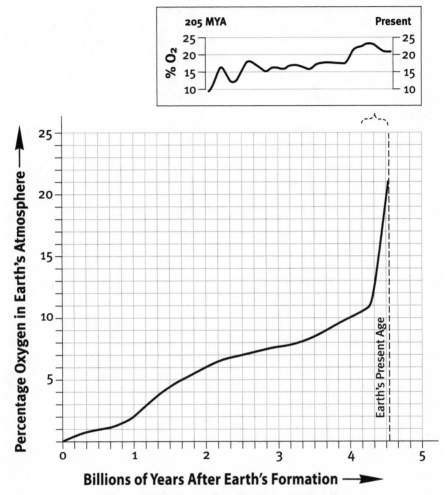

Figure 7.4: Oxygenation History of Earth's Atmosphere
The larger graph shows the increasing abundance of oxygen in Earth's atmosphere over the past 4.1 billion years. The insert shows the increase over the past 200 million years.[17] Note that the explosive emergence of complex, large-bodied animals coincides with the increase of oxygen to a level adequate for the support of such creatures.

complex animals the Cambrian explosion.

Not only did complex animal phyla show up virtually all at once, but so did complete ecologies. Predator-prey relationships, for example, did not develop gradually. They were optimized right from the start of the Cambrian explosion. Furthermore, the most advanced phylum ever to appear on Earth, the chordates (including such vertebrates as jawless fish), shows up at the very base of the Cambrian explosion fossil record.

Contrary to what might be expected from a naturalistic evolutionary perspective, which would anticipate the ongoing appearance of new phyla at a relatively high rate, only five or six new phyla have appeared during the past 540 million years, and about 15 phyla have disappeared. Evolution has proceeded in the opposite direction of what naturalistic models would expect.

The Cambrian explosion does fit RTB's biblical model, which proposes that the Creator worked efficiently and effectively to prepare a home for humanity. A huge array of highly diverse, complex plants and animals living in optimized ecological relationships and densely packing the Earth for slightly more than a half billion years perfectly suits what humanity needs. These life systems loaded Earth's crust with sufficient fossil fuels and other biodeposits to catapult the human race toward technologically advanced civilization.

FILL 'ER UP

Both sedimentation and plate tectonics bury considerable amounts of organic material. Heat, pressure, and time then transform this organic matter into kerogen (high molecular weight tars). More time and heat convert a significant portion of this kerogen into petroleum.[18] Yet more time, plus microbial activity, bring about the degradation of petroleum, turning it into methane (natural gas).[19]

Certain organisms are much more likely than others, upon death and burial, to yield kerogen. The most efficient kerogen producers were the swarms of small-bodied animals that inhabited large shallow seas right after the Cambrian explosion. In the context of providing humanity with the richest possible reserves of fossil hydrocarbons, a fixed period of time had to transpire between the epoch when efficient kerogen producers were dominant on Earth and the appearance of human beings. With too little time, not enough petroleum would have been produced. With too much time, most or all of the petroleum would have degraded into methane.

By itself, the burial of particular organisms and their progressive conversion into kerogen, petroleum, and methane would not be sufficient to produce easily exploitable fossil hydrocarbon reserves. Certain sedimentation processes are needed to lay down porous reservoir rocks. Later, these rocks must be overlaid with fine-grained rocks with low permeability (sealer rocks). Finally, it takes specific tectonic forces to form appropriate caps under which fossil hydrocarbons can collect.[20]

While a long time period was needed for specific sedimentary and tectonic processes to produce appropriate reservoir structures for collecting and storing fossil hydrocarbons, too much time would have led to the destruction of those reservoirs. Other tectonic and erosion processes eventually would have caused the reservoirs to leak. If too much time had transpired, the fossil hydrocarbon reservoirs would have emptied out.

Both methane and kerogen play significant roles in sustaining modern civilization and technology. These substances take a backseat, however, to petroleum, particularly in the plastics industry. While human technology now is advanced enough that it might get by without petroleum, it is doubtful that such technology would have arisen at all without access to large amounts of it.

Human beings indeed arrived at the best possible fossil hydrocarbon moment. Such optimized timing for both the Cambrian explosion and the arrival of humanity would be unexpected in naturalistic models for life but is consistent with what the RTB creation model would anticipate.

RACING THE CLOCK: SPECIATIONS AND EXTINCTIONS

Every species of life gradually changes over time through mutations and natural selection. But a species can evolve into a distinctly different species only if it can naturally select a sufficient number of beneficial mutations before being driven to extinction. What drives a species to extinction is a combination of accumulated harmful mutations, physical changes in its environment, reproductive failures, and competition from other species sharing the same habitat. Changes in the physical environment that would have threatened species with extinction (and still do) include a declining rotation rate of Earth, increasing solar luminosity, changes in tidal patterns, changing chemical composition of Earth's atmosphere, changing biodeposits, nearby supernova eruptions, nearby gamma-ray burst events, asteroid and comet collisions, solar flaring, volcanic eruptions, earthquakes, devastating storms, wildfires, climate cycles, and changes in ocean and air currents.

Therefore, every species races an evolutionary clock. Can a species survive long enough to change through naturally selected beneficial mutations before the onslaught of deleterious mutations and environmental stresses wipes it out? Certain characteristics determine the odds of a particular species' winning the mutational race against extinction:

- Population size
- Average body size

- Average generation span (time between birth and the capacity to give birth)
- Abundance, variety, longevity, and stability of food sources
- Average number of progeny per adult
- Level of parental care and training required for independence
- Duration of parental care and training required for independence
- Complexity of morphology
- Complexity of biochemistry
- Protein-to-body-mass ratio
- Metabolic rate
- Hibernation and aestivation (summer dormancy or torpor) level
- Average life span
- Habitat size
- Ecological diversity of habitat
- Complexity of social structures
- Complexity of symbiotic relationships with other species

Biologists observe that deleterious mutations outnumber beneficial mutations by at least as much as 10,000 to 1, and in some species by as much as 10,000,000 to 1. Thus, the hope of winning any kind of evolutionary race seems extremely dim for the vast majority of species.

Crude models outline the approximate requirements for mutational advance (as opposed to extinction). Direct field observations by biologists reveal significant real-time evolutionary advance for several viral and bacterial species. Among ant and termite species with populations of more than a quadrillion individuals, advance remains debatable. For animal species numbering less than a quadrillion individuals, with average body sizes larger than one centimeter[21] and generation time greater than three months, biologists have yet to observe any significant evolutionary change, other than extinctions.[22]

Significant evolutionary change is here defined as generating a new species that under no circumstances can be made to interbreed with the species from which it arose. Roughly, if the evolutionary limits stated above are valid, biologists should be discovering new bacterial species at a rate that roughly equals or exceeds one per year. And yet, during the past 150 years of research, biologists have failed to observe—in real time—the emergence of even one truly new bacterial species. Such an observational failure implies that the evolutionary limits stated here are not severe enough.

The calculated and observed limits for evolutionary advance imply that invertebrate species cannot, by Darwinian processes alone, give rise to any of the vertebrate species—fish, amphibians, reptiles, birds, and mammals.

Some level of divine intervention must have been involved, whether some interruptive manufacture or manipulation of the natural components of life (transformational miracles) or outright miraculous acts (transcendent miracles) or some combination of the two. With respect to emergence of soulish (birds and mammals) and spiritual characteristics (humans), outright miraculous acts would appear to be required.

When environmental hazards—competition from other species, solar luminosity's continual increase, changes in the chemical composition of Earth's atmosphere, nearby supernova eruptions, deadly gamma-ray bursts, asteroid and comet collisions, solar flares, volcanic eruptions, climate changes, severe storms, and so forth—are factored in, naturalism's odds for progression become even slimmer. All these hazards have contributed significantly to the elimination of species and even of whole phyla from the face of the Earth, well before human activity began to impact the environment.

The RTB creation model offers an explanation for this enigma of the fossil record: Speciation and extinction remained roughly balanced before the appearance of human beings. Then speciation suddenly ended, becoming overwhelmed by extinctions after humans arrived—even apart from human encroachment or abuse. According to Genesis 1, the Creator actively and purposefully built life's diversity as part of the preparation for humanity (in six "days," or eras, of creation), but once humans arrived, He ceased making new kinds of life and no longer replaced extinct life-forms (once the seventh day, or era, of rest came). He assigned humanity the task of managing Earth's living creatures as well as its life-sustaining resources.

"Transitional Forms"

For many decades naturalists have pointed to "transitional forms" in the fossil record as "proof" for their explanation of life's history.[23] The similar bone structures of certain large mammals, in particular whales and horses, through many millions of years supposedly show an apparent progression and thus seem to indicate that such animals evolved by natural process.

Ironically, these so-called best "proofs" may be, in reality, among the worst. Given the evolutionary race against time described above (the smaller the population size, the larger the average body size, the longer the generation time, and the fewer progeny per adult, the less likely a benefit from mutations and the more likely a spiral downward to extinction), horse and whale species would manifest no realistic probability of evolutionary advance and would be expected to survive for no more than several million years before going extinct.

From a biblical perspective, one reason so many apparent transitions appear in the fossil record for whales and horses is that the Creator had a particular time, place, and purpose for each one in the ecosystem. Because these kinds of animals go extinct so rapidly, the fossil record shows frequent replacements, or "transitions," for them. It seems God frequently created new species to replace those that went extinct. Creatures such as cockroaches, with very long extinction times, manifest either no "transitions" or very few. That is, God seldom had to intervene to preserve them.

A naturalistic model would predict a multitude of "transitional" forms among tiny-bodied, simple life-forms, vastly outnumbering those among large-bodied, complex life. RTB's creation model predicts the reverse: far more "transitional" forms for large-bodied, complex life than for tiny-bodied, simple life. The apparent "progressions," or changes, observed in the horse and whale lines more likely reflect the biblical creation accounts' depiction of the step-by-step preparation of the best possible environment, ecosystem, and biodeposits for the focus of the ultimate Earthly creation—humankind.

The existence of numerous "transitional" forms for whales and horses, among other creatures, further suggests that God performed many creative acts rather than just a few along the way. The biblical creation accounts describe God as continually involved and active in creating new species until He created human beings.

DESIGN CONVERGENCE

Striking support for RTB's biblical creation model comes from the observation that species unrelated in the "evolutionary tree" often manifest identical anatomical and physiological features. The limb structures of bats and of flying lemurs, the brain structure for vocalization in hummingbirds, parrots, and songbirds, and the placental anatomy of modern wolves and extinct Tasmanian wolves all show amazing similarities, for example.

Naturalists have attempted to explain such convergence of design as the result of nearly identical environmental, predatory, and competitive pressures on these unrelated species. They propose that natural selection might have shaped these species in similar ways. This explanation poses at least two significant problems. First, design convergence permeates the fossil record. It is bound neither by time nor by circumstance. Given that naturalistic evolution supposedly happens in response to a large number of unpredictable and often dissimilar events, design convergence resulting from natural process should be extremely rare.

Second, design convergence appears in species from radically different habitats facing widely diverse survival stresses. Different habitats imply

(Photo courtesy of Lawrence Migdale/Photo Researchers, Inc.)

(Photo courtesy of David Aubrey/SPL/Photo Researchers, Inc.)

Figure 7.5: Convergent Design

The brain structure of a parrot (top) is virtually identical to that of a chimpanzee. The structure and independent motion of the chameleon's eyes (bottom) are virtually identical to those of the sandlance (a species of fish). From an evolutionary perspective, parrots and chimpanzees are not related, nor are chameleons and sandlances. The fact that these two pairs of unrelated species manifest virtually identical features hints at the work of a Creator who uses optimal design in varying contexts.

different bases for natural selection. An example would be the chameleon, a reptile, and the sandlance, a fish.[24] (See figure 7.5, page 145.) Both have eyes that move independently; when one eye is in motion, the other can remain motionless. Both use the cornea rather than the lens of the eye to focus on objects. Both have skin coverings for their eyes that make them less conspicuous to prey and predators. Both have the same kind of tongue and the same kind of tongue-launching mechanism for snagging prey, and yet these creatures remain far apart on any workable evolutionary chart. The RTB creation model explains convergence as the Creator's efficient use of effective design templates.

The investigation of design convergence by field biologists and paleontologists has barely begun. Likewise, research at the biomolecular level is just starting. Naturalistic models predict that examples of design convergence, both at the organismal and biomolecular level, should prove nonexistent or extremely rare. And they expect the currently established examples to prove flawed.

The RTB creation model, by contrast, predicts that future research will uncover many more examples of design convergence, more examples like that of the chameleon and the sandlance where one can see the convergence of many different intricately complex mechanisms.

LIFE'S CRUCIAL ROLE

Without abundant diverse life on Earth, greenhouse gases couldn't have been converted to rich biodeposits. Without abundant life on the islands and continental landmasses, the quantity of silicates and carbon dioxide converted into carbonates and sand wouldn't have helped launch or sustain civilization.

Certain life-forms, such as vascular plants, are much more effective than other organisms in expediting silicate erosion.[25] This feature proved critical because as Earth aged, silicate erosion by wind and water precipitously dropped (with Earth rotating more slowly) and plate-tectonic activity significantly subsided. Vascular plants were introduced at the just-right time to maintain silicate erosion at a high enough level to ensure that the just-right levels of greenhouse gases would be extracted from Earth's atmosphere. This process perfectly compensated for the ongoing increase in solar luminosity. The quantities and types of vascular plants thereafter introduced needed careful regulation.

In addition, maintenance of all these "just rights" required the support of just-right ecological balances. The life-forms needed for reduction of atmospheric greenhouse gases depended on the existence of other diverse species, all perfectly placed, timed, and proportioned.

The pattern of life observed in the fossil record—which species existed when, where, and for how long—reflects a delicate, step-by-step balancing of nature. The scientific evidence based on observable, objective criteria contradicts the random organization (reversal of chaos) proposed by naturalistic models. According to the past few decades of research, life appeared as soon as Earth was ready for it, no slow-simmering chemical soup ever existed, and life was never very simple.

Another recent discovery shows that life makes a major contribution to Earth's albedo (reflectivity).[26] Life influences the amount of heat and light Earth's surface reflects into outer space, and that amount determines whether or not Earth experiences a runaway freeze-up or a runaway boil-off (see "Climatic Runaways," page 112).

Another new scientific discovery shows that the kinds and quantities of life on Earth's continental landmasses impact the cycling of silica, a substance that buffers soil acidification, regulates atmospheric carbon dioxide, and provides an important nutrient for both marine and terrestrial biota.[27] Each new discovery adds more detail to the picture of an intentional, anticipatory, carefully orchestrated plan—a plan that prepared Earth for life, which prepared Earth for humans.

Some of the most complex creation/evolution issues—questions as to why a Creator would make parasites, why the DNA of higher life-forms is so similar, and why the most advanced species manifest so many examples of apparently "bad" or nonfunctional designs, for example—are addressed in chapter 9. But for now, the momentum of evidence and argument propel the discussion toward the origin and history of humanity.

CHAPTER 8

THE ORIGIN AND HISTORY OF HUMANITY TEST THE RTB CREATION MODEL

More than 100 years before Orson Welles tricked his audience on Halloween Eve, newspaper reporter Richard Adams Locke perpetrated a "hoax so elaborate and fanciful that no one, apparently, could help but believe it."[1] In 1835, knowing that some respected scientists believed in life on the moon, he concocted a tale about a magnificent telescope that showed the moon's most miniscule details.[2]

First, Locke wrote a teaser based on accurate information about a new observatory in South Africa. This piece established his credibility. Then he eased readers into the deception over several installments. One account described "a land that could only be from outer space — idyllic landscapes of brown basalt covered with red flowers, massive green forests, and a blue sea with towering waves and white sand beaches."[3] Another described bizarre creatures. Weeks passed before anyone caught on.

Venerable publications such as *The New York Times* and *The New Yorker* at first gave credence to these amazing "discoveries."[4] It's no wonder, then, that the general public initially accepted these incredible accounts. Authoritative sources confirmed them.

Today, museums, magazines, and zoos tell the story of naturalistic evolution as if all of its claims are unquestionable. Credibility comes from such "proofs" as a moth that adapted to environmental stress by changing its color through natural selection. A recent story about the "human" exhibit in the London Zoo illustrates how firmly evolutionary thinking is entrenched in Western institutions and academia. When children asked why humans were in a cage by the apes, zoo spokesperson Polly Wills said that's exactly the question the zoo wants to answer: "Seeing people in a different environment, among other animals . . . teaches members of the public that the human is just another primate." Chemist Tom Mahoney, who volunteered to join the exhibit, commented, "A lot of people think humans are above other animals. When they see humans as animals, here, it kind of reminds us that we're not that special."[5]

For more than a century, debate has raged over this weighty question: Are humans merely higher primates, or are they something more, something special? A major point of contention and confusion is the usage of the term "human." Anthropologists use it to refer to the broad category of bipedal primates, or hominids, including a number of archaic species distinct from modern humans (see pages 153–157). To avoid any confusion in this book, the term "human" refers only to modern humanity, while "hominid" is used to designate nonhuman bipedal primates.

Debaters about the origin and nature of humans (*Homo sapiens sapiens*) have much to learn from the recent explosion of anthropological and genetic discoveries. It's no exaggeration that more has been discovered about human origins in the past 10 years than in the previous 10,000. Evaluating how RTB's human origins model and its predictions fit with these recent discoveries helps test its viability. While the development of the model remains ongoing, its foundation and forecasts are set forth more fully in *Who Was Adam?*[6] A summary appears in "Human Origins Highlights," page 151. A brief overview of the RTB creation model's distinctive features follows, along with specific scientific evaluations.

WERE HUMANS INEVITABLE?

A central premise of nontheistic models is that the human species arose by strictly natural, unguided steps from a bacterial life-form that sprang into being (or arrived) by strictly natural means on Earth 3.8 billion years ago, just as the London Zoo spokesperson and the chemist asserted. The biblical creation premise contrasts so sharply as to leave little room for ambiguity in discerning which direction the evidence points.

Famed evolutionary biologist Francisco Ayala, an advocate for the assumption that natural selection and mutations can efficiently generate distinctly different species, nevertheless describes the probability that

---

Human Origins Highlights

The RTB creation model for the origin and development of humanity and of global civilization is based on six biblical assertions and premises.

- God created human beings as a deliberate, miraculous creative act.
- Adam and Eve are historical individuals from whom the entire human race descends. This man and woman lived in a particular locale, referred to as the Garden of Eden, somewhere near the juncture of Africa, Asia, and Europe in the relatively recent past (less than 100,000 years ago).
- The human race (*Homo sapiens sapiens*) is qualitatively different from all other animals, including the great apes and the hominids.
- Humanity is Earth's only life-form, past or present, to exhibit a set of spiritual qualities characterized as "the image of God." Manifestations of this spiritual nature include the capacity to evaluate past actions, to contemplate the future, to consider what lies beyond death, to comprehend and attempt to maintain moral and ethical standards, to seek connection with a higher Being or Force, to engage in worship, to express curiosity and creativity about matters far removed from the immediate environment and immediate survival needs, and to seek ultimate hope, purpose, and destiny.
- Humans were gifted from the outset with unique intellectual and physical attributes that would be useful only in a high-tech civilization.
- After the flood of Noah's time, an event that destroyed all humans and all the soulish creatures (the birds, mammals, and perhaps some child-rearing reptiles) associated with them except those on board the ark, humans quickly spread from their locale in or near the Middle East into the rest of Africa, Asia, Europe, Australia, and eventually into the Americas. During this global migration, human population grew rapidly and civilization blossomed.

---

humans (or a similar advanced species capable of developing a high-tech civilization) arose from single-celled organisms as a possibility so small ($10^{-1,000,000}$) that it might as well be zero (roughly the equivalent to the likelihood of winning the California lottery 150,000 consecutive times with the purchase of just one ticket each time).[7] He and other evolutionary biologists agree that natural selection and mutations could have yielded any of a virtually infinite number of outcomes.

Probability calculations by astrophysicists Brandon Carter, John

Barrow, and Frank Tipler produced an even smaller number. Not only does the presumed natural evolution of an intelligent species necessitate a stunningly large number of improbable biological events, but it also demands unlikely changes in the physics, geology, and chemistry of Earth and the solar system. These scientists determined that for an advanced species as technically capable as humanity to arise from a suite of bacterial species in 10 billion years or less, the probability is $10^{-24,000,000}$.[8] Again, the probability for the natural generation of the human species from bacteria or other possible simple life-forms is indistinguishable from zero. To keep this discussion in the realm of reality, it should be noted that a probability as low as 10^{-100} represents a practical impossibility.

The calculations done by Ayala, Carter, Barrow, and Tipler were performed to determine the likelihood that other intelligent species exist in the universe. At the same time, however, they demonstrate the extremely remote probability that humans would exist at all, if nature is wholly responsible.

How Did Humanity Start?

How humans came to exist is a watershed issue for creation/evolution models. If scientists were to prove indisputably that humans emerged from multiple species in various regions rather than from one couple living in one region in the relatively recent past (less than about 100,000 years ago), RTB's biblical model of creation, specifically of human creation, would be dealt a severe blow. The same would be true if researchers were to demonstrate decisively that humans possess no characteristics unaccounted for by intelligence alone. Christian doctrine rests heavily on the belief that the human race is the only species on Earth, past or present, having a spiritual nature inherited from a man and a woman God specially and supernaturally created.

Evolutionary models, by contrast, lean heavily on the belief that humans naturally arose from previously existing species and thus differ from those species only in the degree of their common attributes. Therefore, incontrovertible evidence that humans possess unique attributes or possess no direct genetic link with earlier species would deal evolutionary models a severe blow.

Scientific discoveries about human origins also could falsify young-earth creationist and many theistic evolution models. These models identify Neanderthals, archaic *Homo sapiens*, and *Homo erectus* (referred to below as *H. erectus*) as fully human descendants or predecessors of Adam and Eve. Thus, any evidence that definitively establishes that Neanderthals and these other *Homo* species are unrelated to humans would seriously undermine such models.

Until recently, anthropologists lacked definitive evidence of how humans originated. However, DNA recovery efforts plus extensive new archeological finds, new dating techniques and measurements, and a new abundance of fossils have changed the situation. These recent discoveries offer huge advances toward scientifically validating or falsifying various human origins models, including RTB's.

Inherited DNA?

In sexual reproduction most DNA is recombined, or shuffled like a deck of cards. Exceptions are mitochondrial DNA (mDNA) and parts of the y-chromosomal DNA (yDNA). Everyone inherits mDNA exclusively from his or her mother. Every male inherits yDNA exclusively from his father. Only if that inherited mDNA or yDNA undergoes a mutation can an offspring's mDNA or yDNA be different from the parents'. Thus, measurements of mDNA and yDNA diversity in a population, and of the changes in that DNA over time, serve as an excellent marker for the effects of evolution.

Mitochondrial DNA has been recovered and successfully analyzed from 11 Neanderthal specimens. These 11 specimens span the entire geographical range of Neanderthals (Europe and west Asia) and a significant part of their historical span.[9] This DNA collection shows little variation and no perceptible correlation with either the geographical location of the specimens or their date. In other words, Neanderthal DNA remained remarkably constant throughout the species' range and history.

Geneticists have recovered mDNA from humans dating as far back as 25,000 years ago.[10] The range of diversity for human mDNA does not, in any way, overlap the Neanderthals'. This observation, coupled with marked differences between Neanderthal and human mDNA, establishes beyond reasonable doubt that Neanderthals made no contribution to the human gene pool. They have been eliminated as a possible ancestor for humans.

Dates for *H. erectus* range from 1.8 to 0.5 million years ago with some evidence suggesting dates as recent as about 100,000 years ago. Given the rate at which DNA decays, these dates leave little room for hope of recovering *H. erectus* DNA that's pristine enough for meaningful comparisons with either Neanderthal or human DNA. Nevertheless, *H. erectus* fossils are sufficiently abundant and complete for testing this hominid's role, if any, in the lineage of humanity.

Fossils Record Facts

The discovery that the most ancient fossils for *H. erectus* are indistinguishable from the most recent indicates that *H. erectus* remained static, experiencing no more significant change with respect to time than either

Neanderthals or humans. Because *H. erectus* manifests morphological features radically different from either humans' or Neanderthals' and because all three species experience no observable evolutionary change, it appears highly unlikely that *H. erectus* could be the ancestor of either Neanderthals, archaic *H. sapiens*, or modern humans.

An appeal to earlier hominids in the fossil record as possible links to humanity faces the same problem. All these species manifest morphological features even more radically distinct from humans than does *H. erectus*. Given that earlier hominids would presumably evolve not much more rapidly than *H. erectus*, Neanderthals, or humans (see chapter 7, pages 141–144, for the reasons), it seems impossible that they could naturally undergo the required morphological changes in the time available prior to their extinction.

Wherever extensive fossil evidence exists, it shows that individuals within the different hominid species mature at a much faster rate than do humans. For example, the skull of a *H. erectus* infant (about one year old at the time of death) revealed a brain size about 84 percent of that of an adult.[11] This size compares with 50 percent for a one-year-old human and about 80 percent for a one-year-old modern ape. Likewise, comparisons of Neanderthal skulls from individuals ranging in age at the time of death from six months to young adulthood shows both a much more rapid craniofacial maturation and a much more rapid dental development than humans manifest.[12] Long-lasting childhood and adolescence during which the brain continues to grow and develop appears to be a unique hallmark of the human species. This distinctive provides additional support for the conclusion that hominids did not give rise to human beings.

One Man, One Woman, One Place
The same mDNA evidence that shows humans and Neanderthals experienced no significant evolutionary change and that Neanderthals lack a role in human ancestry also allows scientists to investigate who started the human race, where they came from, and even approximately when. The mDNA evidence establishes that humans descend not from many women from many diverse locations but rather from one woman (or a few women) in a single location.[13]

Likewise, yDNA evidence confirms that humanity descended from one or a few men from the same location.[14] With obvious biblical overtones, geneticists refer to the mDNA ancestor of humanity as mDNA Eve, to the yDNA ancestor as yDNA Adam, and to the single location as the Garden of Eden.

The location of humanity's origin indicated by the mDNA and yDNA analysis is eastern Africa, not the traditional biblical site, Mesopotamia. Identification of the eastern African site, however, is based on the observation that east Africans manifest the greatest genetic diversity of all humanity's ethnic groups—and on the assumption that human migration and human mating is random.

The latter assumption fails the test of both history and observation. Human migration and human mating practices are far from random. For example, both the Bible (Genesis 10–11) and archeological evidence testify to the early, rapid migration of humanity from a single location to diverse distant lands, whereupon the migrating peoples ceased their migration and settled in their chosen destinations.[15]

In most human population groups, people willingly consider a diversity of ethnicities for marriage partners. But, especially in Africa, this was not always the case. For example, 19th-century European explorers to central Africa were surprised to observe tribes of pygmies living adjacent to tribes of extraordinarily tall people. Given that not all human groups migrate and mate like other species, the genetic evidence for human origins simply identifies the single location for the origin of humanity as somewhere near northeastern Africa, which could include any region near the juncture of Asia, Europe, and Africa.

Just as the genetic evidence for the location of humanity's origin remains somewhat ambiguous, so does the biblical evidence. The biblical texts (Genesis 2–4) provide too few geographical clues to pin down Eden's exact location. While a slight majority of Bible scholars prefer the environs of Mesopotamia as the site for Eden, a large minority argue for locations as far afield as Ethiopia, Egypt, Turkey, Persia, and Pakistan.

DNA Similarities

Many evolutionary biologists have argued that the DNA similarity among various species shows that all species are related through evolutionary descent. These scientists further claim that the especially remarkable likeness between human and chimpanzee DNA proves that the two species share a common ancestor in the relatively recent past.

However, as chapter 4 explains, RTB's biblical creation model predicts significantly greater DNA similarity among species than does an evolutionary model. The Bible suggests that God employs the best possible designs and ecological relationships in His creation work. An optimal design for one species often proves optimal for others as well. For example, many metabolic reactions ideal for yeast cells also are perfect for the cells comprising certain human tissues. Therefore, RTB's biblical creation model expects that yeast DNA would share much in common with human DNA.

Morphologically, humans resemble chimpanzees more than they do any other species currently alive on Earth. The average body weight of chimps and humans is similar. So is the structure of all major organs inside the main body cavity. Therefore, from a biblical perspective, it's not at all surprising that human DNA and chimpanzee DNA show remarkable similarity.

New research, however, indicates that the widely advertised 98 to 99 percent similarity between chimpanzee and human DNA is greatly exaggerated. Such claims are based on small segments of the human and chimpanzee genomes where common sense dictates that the similarities would be the greatest. While comparisons between the complete human genome and the complete chimpanzee genome have only recently begun, the most complete comparisons performed thus far indicate that the degree of similarity is more like 85 to 90 percent.[16]

Dissimilarity is especially pronounced for gene expression patterns that govern brain structure and activity. The human brain, unlike the brains of chimpanzees or any other species that has ever lived, possesses structures to sustain spiritual activity, meditation, analysis, mathematics, logic, complex language development, and communication. The gene expression patterns responsible for these structures are unique to humans.[17]

The RTB creation model predicts that as geneticists look deeper into the genomes of the great apes and the hominids that preceded humanity (see "How Did Humanity Start?" page 152), they will continue to confirm that the human species is genetically distinct, not linked through natural evolutionary descent to other primates. The RTB creation model predicts that future genetic research will attest humanity's uniqueness—a special creation in whom the Creator made appropriate use of similar or identical genetic designs He already optimized for other species.

Why So Many Different Hominids?

The Bible makes no direct mention of the hominids that preceded humanity in the fossil record. Such an omission is consistent with a pattern of avoiding vocabulary or references to natural phenomena that only a few biblical readers throughout all the centuries would be familiar with. (That may be one reason why dinosaurs, fundamental particles, galaxies, and penguins receive no direct mention in Scripture, aside from practical concerns for the documents' length.)

Scripture, therefore, does not directly address the question of why God created so many different hominid species. However, some recent ecological research points to some possible reasons for God's creating these species. One is humanity's destructive impact on the environment. There could be other reasons as well.

(continued)

The Bible does address man's responsibility for the natural realm[18] and warns of the negative impact of man's sin upon other creatures.[19] Given that God gave soulish animals (birds and mammals) the desire to interact with humans, the evil that humanity manifests could have had an utterly devastating impact on such species. According to Genesis 9:2, God took protective action by bringing "the fear and dread of [humanity] ... upon all the beasts of the earth and all the birds of the air."

From a biblical perspective, because God possesses complete knowledge of the future, He would have known before creating any hominid that the future human race would rebel against His authority, becoming selfish and dangerously neglectful of one another and their physical environment. The number of bird and mammal species driven to extinction gives sad testimony to humanity's abuse of the environment. Of the 15,000 to 20,000 bird species present on Earth at the time of humanity's origin, only about 9,000 still remain. Of approximately 8,000 land mammal species, only about 4,000 still remain.[20]

Humans have devastated the very creatures they are most dependent upon for their livelihood and quality of life. Perhaps if God had not created a series of progressively more advanced hominids, human impact could have been even more devastating. Evidence to support this conclusion comes from large mammal extinction studies. In Africa, where several hominid species predated humanity, the extinction rate for large mammals during the human occupation period is 14 percent. In North and South America and Australia, where no such hominids preceded humans, the large-mammal extinction rate during the human occupation period stands at 73, 79, and 86 percent, respectively.[21]

The fossil record reveals a sequence of hominids in such places as Africa that spanned several million years, with each successive species slightly more capable in hunting birds and mammals than the previous one. This slowly increasing exposure to gradually improved predation skills may have allowed birds and mammals to adapt step-by-step for the shock of a sinful, superior predator.

Humanity's Birth Date

During the past decade anthropologists, archeologists, geneticists, and geochemists have developed a diverse collection of new tools for dating the origin of humanity. As a result, humanity's birth date is no longer a terribly vague mystery.

DNA determination. The date determined by geneticists for the first yDNA man is roughly 42,000 to 60,000 years ago.[22] The date for the first mDNA woman has reflected a much wider error bar, 170,000 ± 50,000 years ago.[23] This mDNA date assumes universal homoplasmy (that all humans possess only one type of mDNA). However, according to more recent studies, 10 to 20 percent of the human population possesses two types of mDNA (heteroplasmy), and nearly 1 percent has three types (triplasmy).[24]

Calculations based on these new findings place the date for the first mDNA woman closer to 50,000 years ago, in line with the yDNA date.[25] It also corresponds with the biblical date for the creation of Adam and Eve based on reasonable calibration of the Genesis genealogies.[26] The dates for the explosive emergence of advanced art, advanced tools, complex language, clothing, and jewelry corroborate this timing for the first human pair.

Family records. Bible scholars acknowledge that gaps exist in the biblical genealogies. They disagree, however, as to how many and how wide. The genealogies in Genesis 5 and 11 trace human history from the first man, Adam, to Abraham, the father of both the Jews and the Arabs. Historical records, biblical and extra-biblical, independently establish that Abraham lived about 4,000 years ago.

An accurate carbon-14 date places the breaking of the Bering land bridge (the land bridge that once connected Asia to North America) at about 11,000 years ago.[27] Genesis 10:25 declares that the world was divided in the time of Peleg, one of the patriarchs mentioned near the midpoint in the Genesis 11 genealogy. Given that life spans declined geometrically from around 950 years for the earlier patriarchs to about 120 years for the later, and given that the life span recorded for each patriarch listed in the Genesis 5 and 11 genealogies is proportional to the actual passage of time (a reasonable though unproved assumption), extrapolation from the fixed dates for Abraham and Peleg would indicate that Adam lived approximately 50,000 years ago.

Tool kits. Up until about 50,000 years ago the most advanced tools constructed by hominids were flake fragments (useful for scraping, for example) created by striking a core rock with a hammer rock.[28] Between 50,000 and 40,000 years ago, however, a quantum leap in technology occurred.[29] Early humans made use of wood, leather, sinews, bones, ivory, and stones for the construction of tools. They found ways to carve and polish raw materials into sophisticated implements that included axes, hammers, knives, fishhooks, harpoons, awls, needles, and shovels.

A cultural explosion. A technology big bang coincided with a burst of clothing,[30] jewelry, and other cultural developments. No evidence currently exists for the manufacture or use of any of these trappings by hominids. For humans, however, use of jewelry and clothing exploded into the archeological record all at once. At the most ancient human sites the quantity of jewelry outweighs the quantity of tools and the quantity of seashells used for making jewelry outweighs the evidence of shell fish used for food.[31] Similarly, archeological research reveals other coincidental events: the sudden and widespread appearance of advanced art, musical instruments, complex language, and religious ceremonies and practices.[32]

All of these cultural explosions date to about 40,000 years ago, timing consistent with a human origin date roughly 50,000 years ago. (A significant period of population growth and societal development was necessary for deposition of any archeological evidence that modern-day scientists could discover.)

Over-Endowed Humans

Human beings are vastly over-endowed for a hunter-gatherer or simple agricultural lifestyle. For tens of thousands of years humanity possessed innate talents beyond obvious purpose or advantage. Though many examples could be cited, dexterity, intellectual capacity, and sex drive are among those that set *Homo sapiens sapiens* apart from all other hominids and animals.

These extra capacities are costly. They take energy, time, muscles, nerves, and blood flow to support. Such resources could otherwise be channeled to features that would provide immediate survival advantages. From a naturalistic perspective there is no mechanism for generating abilities that offer no immediate advantages to the species.

Dexterity. The design and agility of the human hand certainly gave the human race an early survival advantage. Humans could craft more elegant tools and weapons than hominids. However, the ability to type faster than 100 words per minute offered no obvious survival advantage until it met the corporate world. Likewise, the ability to play a Liszt concerto had no benefit until the invention of the modern piano.

Intellectual capacity. The intelligence quotient of the human brain gave *Homo sapiens sapiens* a huge survival advantage, providing them the ability to invent new implements for hunting, agriculture, cooking, and building. And yet, not until the 20th century was any use found for the phenomenal capacity of the human brain to tackle higher mathematical functions such as nonlinear tensor calculus, relativistic quantum theory, and higher-dimensional geometry.

Humans pay a price for their ability to engage in higher mathematics. The main reason humans are weaker and slower than similar-sized animals is that 35 percent of the entire blood flow in the human body serves the brain. As it is, so much brain structure is needed to support the capacity for higher mathematics—as well as meditation, analysis, prayer, logic, and complex languages—that brain structures, which in other animals are devoted to supporting more muscles and more acute senses of smell, hearing, touch, and/or taste, must be sacrificed. (The option of a bigger human brain is not available because the human brain already requires so much blood flow as to be in danger of overheating.)

An active sex drive. Many biologists have pointed out that compared to other mammals, humans are dramatically oversexed. Their drive is unusually strong and enduring. Whereas females of other mammal species are receptive to males only at a certain season, in some cases for just a few days each year, human females mate throughout the year. This tremendous capacity for sex explains how the human race was able to multiply from one couple to 6 billion people in a relatively short time period. Genesis 1 and 9 lay out the rapid multiplication of humanity as God's specific goal.

An extraordinary desire for sex is particularly critical in a high-tech society. Both affluence and technology provide powerful diversions from human reproduction.[33] If not for an exceptional sex drive, the human species would be challenged to survive in such an environment.

An assessment of the evidence. Some biologists have proposed that certain currently useless byproducts of evolutionary change might remain long enough to become useful for a future challenge or application. However, naturalistic evolutionary principles would permit such survival only if the cost to the individuals of a species for carrying such a load were relatively low. In the examples above the cost is high.

Many other endowments also equip humanity for maximum performance in a high-tech environment. Unlike any other species of life, humans seem to have been equipped in advance for a future role far different from the one they fulfilled when they first appeared.

From a naturalistic perspective, perhaps one, but not all, of humanity's over-endowments might appear as a random genetic accident in a single individual. However, the evolutionary forces and mechanisms that characterize all naturalistic models would quickly eliminate such immediately unusable but costly accidents. On the other hand, such extensive equipping of human beings in advance for later needs and the preservation of such capacities points to One who has foresight and desires to prepare humans for their future global technological endeavors (see chapter 4, page 84).

FOR A LIMITED TIME ONLY

The RTB creation model predicts specific dates for the end of all life as well as for the beginning. The difference between the two dates, at least for humans, should be relatively small, according to the Bible: some tens of thousands of years, as opposed to millions or billions. What does science say about a rapidly closing time window for humanity?

Greenhouse gases in Earth's atmosphere currently warm the planet's surface temperatures by 60°F (33°C) above what they otherwise would be. Such warming makes life, including human life, possible on Earth. The sun, however, grows progressively brighter as it continues to burn or fuse hydrogen into helium. The extra potentially deadly solar heat Earth will receive in the future could conceivably be compensated for by the removal of the just-right amounts of greenhouse gases from the atmosphere as in the past (see pages 133–134). However, this process cannot persist for much longer.

The only significant greenhouse gases left in Earth's atmosphere today are water vapor and carbon dioxide. Reducing water vapor in the atmosphere means reducing the amount of rain that condenses out of the atmosphere. Water vapor reduction sufficient to compensate for the future brightening of the sun would turn all the continents and islands into parched deserts.

Carbon dioxide plays a much bigger greenhouse role. To compensate adequately for a brightening sun, the amount of carbon dioxide in the atmosphere must be reduced by a large proportion. The current carbon dioxide level is 375 parts per million. As this level declines, so does photosynthetic plant production. When the atmospheric carbon dioxide level falls below about 225 parts per million, all photosynthetic life dies, and the death of most (if not all) animal life quickly follows.

For the past 3 billion years or so, greenhouse gases in Earth's atmosphere have been steadily reduced so that life can remain abundant on Earth despite a brightening sun. However, continued reduction of greenhouse gases extends the time window for life by only another 0.02 billion years. The first life to disappear will be the large advanced animals. A nearby supernova eruption, a climatic perturbation, war, a social or environmental upheaval, a mass extinction of supporting species, a declining birth rate, or the genetic accumulation of negative mutations could drive humans to extinction in just tens of thousands of years. The last to go extinct will be bacteria. Earth should one day become sterile.

The timing of humanity's arrival near the inevitable end of life on Earth is either a dreadful but meaningless fact or part of the Creator's plan. By coming at the end of life's history on Earth, human beings reap the benefit of nearly 4 billion years' worth of biodeposits. A look around the cities and transportation arteries that link them reveals that the raw materials necessary for their construction came from biodeposits—concrete, iron, zinc, chromium, molybdenum, limestone, marble, bricks, mortar, asphalt, timber, paper, plastics, and so forth. Nearly all the energy that drives civilization comes from biodeposits such as coal, oil, natural gas, wood, and kerogen. In addition, biodeposits supply nearly all the fertilizers (such as phosphates and nitrates) that support agricultural production. And, more importantly, human life continues beyond Earth's demise according to the RTB creation model.

THE CREATOR'S FINGERPRINTS

Rather than being animals to observe in a zoo, human beings (*Homo sapiens sapiens*) belong in a category all their own. The DNA and cultural evidence, the fossil record, the timing of humanity (among many other evidences) all point toward the plausibility of creation. Earth's history from beginning to end reflects a Creator's care for humankind. Getting beyond the presuppositions and superficial inquiry of many so-called experts to examine the accumulating scientific evidence leads to some stunning insights. Testing the evidence for human origins yields a potent case for creation.

However, before RTB's biblical creation model can gain respect among naturalists and others, it must address the most difficult *why* questions concerning a Creator's intentions. Answers to these challenges and others provide yet another means for putting a spectrum of creation/evolution models to the test.

CHAPTER 9

THE *WHY* CHALLENGE

*I*f a failed educational system and/or "dumbbell mentality" were chiefly responsible for Welles' success in pulling off a Halloween scare, *why* were brilliant people among those who panicked and ordinary folks among those who sat back and enjoyed the drama? If fear of flying truly reflects fear of death, why does a terrified flyer calmly drive the Los Angeles freeways? If so-called "acts of God" represent acts of judgment upon godless people, why are God-loving people among those severely affected by such events?

Why questions probe plausibility. *Why* questions ferret out weaknesses—and strengths—in an explanation. *Why* questions can lead to breakthroughs in understanding. Some of the greatest discoveries of astronomy, medicine, physics, biology, anthropology, criminology, and other disciplines of knowledge began with *why* questions. Such questions may play a significant role in resolving evolution/creation controversies.

At the kick-off banquet for a 2004 creation conference at Biola University, Cornell biology professor William Provine, a frequent debater for evolution, dared the gathering of creation proponents to stop avoiding the *why* questions posed by evolutionists. He charged that creationists, the Intelligent Design movement (IDM), and the notion of a divine Creator will never be taken seriously by the scientific community until believers in creation can explain such things as why an intelligent Designer would cause multiple extinctions of life and why human DNA is so similar to chimpanzee DNA. The human-chimp DNA problem, Provine pointed out, is particularly symptomatic of the IDM spokespersons' aversion to dealing with the issue of human origins.

Evolutionists, too, face a list of *why* questions they seem loathe to address. For example, if life originates through strictly natural processes, why have chemists failed to produce life or even DNA in the lab from inorganic compounds? Why have they failed to suggest reasonable chemical pathways whereby such outcomes could occur? Also, if humans naturally evolved from lower life-forms, why do they possess so many totally unique features and abilities, and why are so many of these features (such as mathematical and musical capabilities) detrimental to basic survival?

The failure to grapple with challenging *why* questions amplifies polarization and frustrates the possibility for any meaningful resolution of evolution/ creation controversies. One popular Web site refers to the persistent impasse as "The Eternal Debate."[1] Not only does avoiding critical *why* questions guarantee unending conflict, but it also seriously damages, as this chapter discusses, both science education and the public's enthusiasm for science.

THE RTB CREATION MODEL TAKES UP THE GAUNTLET

The questions posed by Provine and other skeptics of creation have already played a helpful role in expanding and refining the RTB creation model. Some of these *why* questions will be addressed in this chapter:

- Why would a loving Creator make carnivores and parasites?
- Why would an intelligent, supernatural Creator make "junk DNA"?
- Why do "bad designs" exist in nature?
- Why would a caring Creator expose all His creatures to so many destructive "acts of God," such as earthquakes, hurricanes, tornadoes, volcanic eruptions, floods, drought, and wildfires?

Other important questions, addressed briefly in chapters 5–8, that must be answered in depth by any creation/evolution model are these:

- Why would an all-powerful Creator take 10 billion years to prepare the cosmos for life?
- Why would an all-powerful Creator take 4+ billion years to prepare Earth for human life?
- Why must humans and other animals suffer from cancer and other diseases?
- Why would a caring, all-powerful Creator cause millions of generations of plants and animals to die on His way to making humans?
- Why would a Creator bother to make so many galaxies, stars,

planets, moons, and asteroids, and so much dust, if human life needs just one star and one planet?
- Why would a Creator make so many bacteria, trilobites, and dinosaurs?
- Why would a Creator make so many different hominid species before creating humans?
- If everything has a beginning (and end), why doesn't God?
- Why don't more of the brightest and best-educated scientists recognize and acknowledge the evidence for a Creator?
- Why do so many intelligent Christians (among others) embrace nonsense science?

More thorough discussion of these latter issues can be found in other RTB books.[2] The questions that are addressed in this chapter admittedly deserve a more thorough treatment than space here allows.

WHY CARNIVORES AND PARASITES?

The often romanticized vision of Adam and Eve and all the animals dwelling in an earthly paradise free of death, pain, decay, and suffering—suddenly ruined by Adam's bite from a forbidden piece of fruit—falls short of both biblical and scientific reality. While the Bible does teach that Adam's rebellion intensified pain and trouble for both his own and other earthly species and visited death upon humanity, it does not say or suggest that no pain or death or other "unpleasantness" existed prior to this time.[3]

Biblical accounts of nature, particularly those in Job and Psalms, describe various species of life as fulfilling vital roles for other species in ecological systems. As part of God's provision and care for life, He designed elegant ecologies that include, yet minimize, pain and death (apart from human interference).[4]

God's optimal ecological designs establish circles of life that both prevent species from over-consumption and ensure that the nutrients life needs cycle efficiently throughout the environment. Through such protection of the food supply and nutrient base, suffering and death are minimized for all species.

Nature observers note the strange fact that plants vastly outproduce what they need for survival. This overproduction provides for the existence of animals. Herbivores consume the overage. As they consume and process this food, they both transform and transport the nutrients obtained from plant matter to other places and uses in the environment. If it were not for herbivores' feeding on them, plants would deplete their particular nutrient base to a point that threatens their existence, population level, habitat spread, genetic vitality, and/or health. Herbivores also benefit plant populations by spreading seeds over a wide area, sometimes introducing them to new ecosystems.

Just as plants need herbivores to maintain their vitality, so also herbivores need carnivores. Unlike human hunters, other carnivores focus their hunting energies on the sick, the injured, the unwary, or the weak for their food supply. By removing these individuals from herbivore flocks or herds, carnivores alleviate herbivore suffering and prevent herbivore populations from becoming dangerously diseased and genetically weakened. They also stop herbivores from exhausting their food resources and starving to death. The current overpopulation of elephants reported in certain parts of Africa demonstrates how serious the problem can become when herbivores lack adequate carnivore predation.

> Observation and research in areas where large elephant populations have been allowed to naturally grow have shown that elephants are capable of turning woodland into grassland. This means that the species that thrive in wooded areas would be lost from the ecosystem. Wildlife conservationists are then faced with a dilemma—do they conserve elephants at the cost of numerous other species, and biodiversity in general, or do they artificially limit elephant numbers?[5]

Carnivores offer an additional benefit to humans. Genesis 1 says that God created two kinds of advanced land mammals to cohabit with humans—land mammals that are wild, that is, difficult to tame, and land mammals that are not so wild. These easy-to-tame land mammals, the herbivores, make excellent animals for food production (and other agricultural purposes) but rather poor pets. Such mammals can't be housebroken and because of their constant need to eat and ruminate are less available to relate to and entertain humans. The hard-to-tame land mammals, carnivores such as cats and dogs, by contrast make excellent household pets (but poor agricultural producers). According to the Bible, God made both kinds of land mammals for humanity's well-being and enjoyment.

Unlike most carnivores, parasites do not kill their "prey" quickly or immediately. But they, too, perform useful purposes. Some parasites, including certain species of bacteria, indirectly protect their hosts from calamity. For example, humans who suffer the unpleasantness of frequent bouts of parasite-induced diarrhea experience a significantly lower risk of contracting colon cancer. Some parasites distract their hosts from feeding or encourage them to leave a particular habitat and thereby give species lower on the food chain a reprieve. Parasites, like carnivores, may help control their host species from overpopulation (see "Why Would a Good God Create Parasites?" below) or pressure them toward healthier behaviors.

Why Would a Good God Create Parasites?

Research may never fully reveal all the good that parasites accomplish in the balance of nature. But a true story illustrates some of the ways parasites can benefit Earth's life and resources in much the same way carnivores benefit herbivores:[6]

Famed Harvard anatomy professor Dr. Étienne Léopold Trouvelot took up the study of exotic insects as a hobby. One afternoon in 1868 a few prized gypsy moths obtained from Europe escaped from his home laboratory. Unchecked by any local predators, this moth species' population ballooned to pandemic proportions. Within several years, deciduous forests across New England, then over most of the eastern United States, were stripped of every leaf. Not only were these forests virtually wiped out, but hundreds of species dependent on them suffered catastrophic population declines also, including the gypsy moths themselves.

For the first few decades of these infestations, nothing could slow or stop the moths — except the lack of food once they had stripped the forest foliage bare. After each destructive episode, the forests took decades to recover, but when the trees did come back, so did the gypsy moths. Each cycle produced a progressively weaker gene pool for all the plant and animal species involved.

When local carnivores, primarily birds and mice, finally adapted to the gypsy moths as a new food source, the degree of devastation decreased some. But it did not end.

Some significant headway was made toward solving the problem when researchers introduced a European virus specific to the gypsy moth. But not until 1989 did the destructive cycles finally near an end. That's when scientists brought in a second parasite. Another European pathogen, this time a fungus, came to the rescue. It took several carnivores and at least two parasite species feeding on gypsy moths to ensure that North American deciduous forests will remain extensive enough and healthy enough to sustain hundreds of species, including gypsy moths, with an optimal quality of life.

This story provides an example of how the lack of appropriate parasites resulted in loss for everybody (all species in the habitat). It shows that with an adequate number and diversity of parasites everybody wins, including the species that the parasites attack. The case of the gypsy moth demonstrates that where fairly complete ecological knowledge and understanding is available, the existence of well-designed parasites really does comport with the plan of a caring, powerful Creator.

From an evolutionary perspective, optimal ecological relationships —plants, herbivores, carnivores, and parasites in well-balanced relationships, with no species dominating the habitat to the detriment of others—would be unexpected. Either they would not develop at all, or they would develop very slowly and haltingly. A biblical perspective anticipates creation of life with optimal ecological relationships right from the start. So far, fossil record research on mass speciation events confirms that these complex, optimal ecologies arose virtually intact. The RTB creation model predicts that future research into fossil record speciation events will provide further validation.

WHY JUNK DNA?

In complex plants and animals, only 5 to 20 percent of an individual's DNA carries code for making proteins. In humans, only about 3 percent of the DNA codes for proteins. Geneticists have long referred to the DNA that does not code for protein production as "junk" DNA. From an evolutionary perspective, junk DNA seems expected and explainable. Undirected biochemical processes and random molecular events would likely transform functional DNA segments into useless artifacts. These obsolete components tag along from generation to generation in the genome (an organism's total nuclear DNA content) due to their physical attachment to the functional strands of the DNA.[7]

For decades this explanation satisfied curiosity, and many scientists considered it powerful evidence for naturalistic evolution.[8] So, when identical segments of junk DNA appeared in a wide range of species that from an evolutionary perspective are related to one another—often in the same genome location—evolutionists drew what they considered an obvious conclusion: respective junk DNA segments arose prior to these organisms' divergence from a shared evolutionary ancestor.[9]

The assumption that the non-protein-coding part of the genome served no purpose caused researchers to abandon study of its features for nearly three decades. Then a team of physicists made an observation that revived interest. They noticed that the quantity of "junk" in a species' genome was proportional to that species' degree of advancement.

The physicists decided to perform a computer analysis, and in 1994 they published their results. They found that what had long been labeled junk DNA carries the same complex patterns of communication found in human speech.[10] In fact, they found that the junk DNA had an even higher linguistic complexity than did the protein-coding DNA. This breakthrough discovery drew teams of geneticists worldwide into a veritable frenzy to uncover the hidden designs and functions of the portion of DNA once thought useless.

This flurry of research has revealed five kinds of noncoding (for proteins) DNA, and each kind plays an important role in the vitality and function of the organisms in which they reside (see appendix D, "Functional Roles of 'Junk' DNA," pages 221–222). However, much of this research is so recent that a considerable quantity of what has long been termed "junk" has yet to be studied.

The RTB creation model anticipates that future research into the remaining "junk" DNA will provide further evidence of purpose and design. It does not, however, propose that all DNA must serve a functional purpose. The RTB creation model acknowledges that the optimal DNA designs present at the time God created a species will, thereafter, gradually degrade as a consequence of natural mutations. Therefore, depending on how long a particular species has existed, a small amount of real junk is expected to be present in the DNA.

WHY BAD DESIGNS IN NATURE?

For more than a century, evolutionists have argued that if a caring, powerful God were behind Earth's life, organisms would be free of obviously inferior designs. That these bad designs are most apparent in the late-arriving, large, complex animal species would seem to indicate that bad designs accumulated as a result of evolutionary processes.

Perhaps the most famous examples of so-called "bad designs" in nature are those identified by the late Stephen Jay Gould in his book *The Panda's Thumb*.[11] As the title suggests, the centerpiece of Gould's case for bad design is the "ineffective" thumb of the giant panda. He viewed it as a clumsy adaptation of wrist bone material, not the work of a divine Designer.

While rebuttals to Gould's argument have been published since the mid-1980s,[12] a newer study offers the most rigorous response. Six Japanese biologists used three-dimensional computer axial tomography and magnetic resonance imaging (CAT and MRI scans) to determine that "the radial sesamoid bone and accessory carpal bone form a double pincer-like apparatus in the medial and lateral sides of the hand, respectively, enabling the panda to manipulate objects with great dexterity."[13] In the close of their paper the Japanese biologists concluded, "The hand of the giant panda has a much more refined grasping mechanism than has been suggested in previous morphological models."[14] Their conclusions were confirmed by field observations of three pandas. Those studies showed the wrist flexion and manipulation of the double-pincer capacities are essential aspects of the panda's specialized food gathering and feeding.

Not too long ago, surgeons doing abdominal surgery would routinely remove the appendix, whether it was inflamed or not, on the presumption that it was a useless byproduct of humanity's evolutionary history. With the discovery that the appendix plays an important role in the human immune system, the practice has stopped. Likewise, textbooks on anatomy once claimed that the "tailbone" at the base of the human spine was a useless byproduct of humanity's descent from long-tailed primates. As a result of research into the engineering dynamics of the human spine, anatomists now recognize that the human tailbone and, in fact, all the bones of the human spine and its S-shape are exquisitely designed to facilitate bipedal walking, standing, and load carrying.

"As useless as nipples on a man" may be a familiar cliché, but as it turns out, male nipples are not entirely useless. For one thing, the manufacture of the human body and the bodies of all animals requires seam points in the skin. Male nipples appear to serve as convenient seam points for the Manufacturer.

More than just convenient seam points, though, male nipples represent efficiency. An observed attribute of the Creator, noted both in the record of nature and the words of the Bible, is His economy of miracles—only what's needed to accomplish His purpose. This economy implies that the morphological differences between men and women would be few, not many. Nipples with "low" functionality for men permit a more common genetic template for men and women than if God had created different male seam points and placed them in different locations.

Another known function of male nipples is the role they play in sex. As with women, men's nipples are densely packed with nerves. The sensitivity these nerves produce amplifies pleasure and makes this area of the male body an erogenous zone. That is, the nipples on both men and women are so sensitive to touch that they fulfill a significant role in sexual stimulation.

Questions about the design of the panda's thumb, the human appendix and tailbone, and male nipples should caution scientists against jumping too quickly to an evolutionary conclusion whenever some aspect of anatomy seems superfluous. The RTB creation model anticipates that future research into the anatomy of complex animal structures will reveal increasing, rather than decreasing, evidence for exquisite design and functionality.

Why Hurricanes, Earthquakes, Volcanoes, Wildfires, and Ice Ages?

No doubt hurricanes, tornadoes, volcanoes, earthquakes, wildfires, ice ages, floods, and droughts cause inestimable damage and untold suffering to plants, animals, and humans. Though it cannot and does not diminish the pain and loss, the fact is that even more damage and suffering would result if none of these "acts of God" ever occurred.

Hurricanes and Tornadoes?

It would be possible for a Creator to get rid of hurricanes and tornadoes, for instance. Wind velocities depend on Earth's rotation rate. The more slowly Earth rotates, the fewer, shorter, and less intense its tornadoes and hurricanes. If Earth's rotation rate slowed to 26 hours per day, no hurricanes or tornadoes would ever occur. However, a rotation rate that slow would make for significantly colder nighttime temperatures and significantly warmer daytime temperatures all across Earth's surface. Such temperature extremes between day and night would cause even greater suffering and death than do the current number, intensity, and duration of hurricanes and tornadoes. A slower rotation rate would also result in less rainfall and more sporadically distributed rainfall on continental landmasses.

Already Earth rotates more slowly than it did in the past. As noted in chapter 5, the moon acts as a gravitational brake on Earth's rotation rate. Bands in 400-million-year-old coral reefs show that at the time these corals formed, Earth rotated more than 420 times (days) in a year, each spin taking less than 21 hours.

These 21-hour days spawned enormously more destructive hurricanes and tornadoes. Placing humanity on Earth when the rotation rate had slowed to 24 hours meant that the Creator timed the human era to correspond with the ideal hurricane and tornado era in geologic history—another piece of evidence that the timing of humanity's advent was planned rather than accidental.

The argument for creation goes beyond the notion that while hurricanes and tornadoes are bad, the alternatives are worse. Hurricanes and tornadoes actually serve several good purposes. For example, hurricanes significantly increase chlorophyll concentrations along continental shelves. Semiregular chlorophyll enrichment enhances both the biomass and biodiversity of the all-important life-forms that reside on these continental shelves.[15]

Hurricanes linger over oceans far longer than over land. Their powerful winds lift huge quantities of sea-salt aerosols from the oceans. These aerosols make up a large fraction of cloud nuclei, which in turn play a critical role in raindrop formation.[16] Thus, hurricanes (and, to a lesser degree, tornadoes) ensure that enough rain falls from the atmosphere to support a large and diverse population of land life.

These aerosols and the clouds that form from them also efficiently

Figure 9.1:
Satellite Image of Hurricane Katrina, Which Devastated New Orleans in 2005
(Photo courtesy of NOAA/Photo Researchers, Inc.)

scatter solar radiation. So hurricanes also fulfill a life-essential role as Earth's thermostat.[17] When the tropical oceans get too hot, they generate hurricanes. The sea-salt aerosols produced by hurricanes cool the tropical oceans to a benign temperature. To a lesser degree tornadoes act as thermostats also, cooling certain continental landmasses that have become too hot.

Earthquakes and Volcanoes?

As noted previously (pages 134–138), plate-tectonic activity, which gives rise to earthquakes, plays a critical role in building islands and continents, in compensating for the sun's increasing luminosity, and in maintaining life-essential chemical cycles. It also provides an ongoing supply of nutrients to surface soils.

The maintenance of these life-essential processes early in Earth's history required much greater tectonic upheaval. Today, the level of activity is only about a fifth of what it was when Earth's first life came on the scene. As with hurricanes and tornadoes, scientists note that the human race appears on Earth at the ideal tectonic moment. Earthquake activity today is high enough to sustain adequate levels of various surface nutrients but low enough to allow for a global, high-tech civilization.

Ice Ages?

Scientists are just beginning to understand the life benefit of ice ages. For some time, however, geologists have noted that the large, fast-moving glaciers predominant during ice ages contributed to the formation of many of Earth's richest ore deposits. Geographers observe that ice ages, and the resultant glacial sculpting of Earth's crust, are responsible for carving excellent harbors, fertile valleys, and gorgeous lakes on high-latitude landmasses.

Wildfires?

No one needs a scientist to prove that life on Earth would be severely impacted, both in quantity and in diversity, if wildfires were more frequent and widespread. Scientists have been helpful, however, in identifying the primary factors that determine how fires start and burn: (1) the quantity of atmospheric oxygen, and (2) the electric discharge rate (rate of lightning strikes). If these were greater by even a small percentage, fires would seriously limit the level of human civilization and technology.

Figure 9.2: Wildfires
Large forest fires such as the one pictured here infuse significant quantities of charcoal into the soil, remove growth inhibitors, and stimulate the production of nitrogen-fixating microbes. To maintain the quality of forest soils, it is best that old forests burn about once every 20 to 60 years.

(Photo courtesy of David R. Frazier/Photo Researchers, Inc.)

New soil science studies reveal that humanity would be in similarly serious trouble without *enough* forest and grass fires.[18] First, fires get rid of certain growth inhibitors. Anyone walking through an old-growth forest notices the accumulation of dead vegetation on the forest floor. Burning off this organic litter enhances germination by giving seeds and seedlings greater access to the mineral soil beneath. Old forests also accumulate certain plant- and microbe-suppressing agents. Burning stimulates essential microbial activities, such as nitrification of the soil. The lightning that starts many fires contributes by generating nitrogen fixation from the atmosphere.

A by-product of forest and grass fires is charcoal, which benefits the soil by absorbing tannins and other plant- and microbe-inhibiting chemicals that have settled there. During and after a fire, charcoal breaks down into fine dust and ash, which is easily transported by wind and water to areas adjoining the burn area, especially to valleys. This dust and ash, relatively chemically inert, greatly enhance the soil's water retention capacity. They can transform even sandy soil into a clay-like material.[19] Further, this dust and ash are a significant source, perhaps the most significant source, for the development of new wetlands. They may also play a role in the development of peat bogs and, thus, coal formation.

Like charcoal filters used in water purification systems, soil charcoal benefits dwindle with time. Studies of Swedish forests indicate that the benefits of charcoal drop to one-eighth their original level in 100 years. After 200 years, no measurable benefit remains. Studies in American forests demonstrate that fires every 20 years, for example, do much less damage than fires every 150 years. These findings have led researchers to estimate that Earth's biomass and biodiversity are maximized if forests and grasslands burn every 20 to 100 years—precisely the natural rate ecologists measure for much of the planet.

In response to the question, "Could not an all-powerful Creator alter cosmic and terrestrial physics so that humans could exist without such things as hurricanes, earthquakes, volcanoes, wildfires, and ice ages?" science answers *yes*, but not without disturbing Earth's system in ways that would thwart several of His purposes for creating the universe and Earth and life and human life in the first place. As stated in chapter 4 (pages 81–82), the Bible says that God will create a realm where no such phenomena ever occur, but only *after* evil is permanently and finally done away with.

WHY QUESTIONS FOR EVOLUTIONISTS AND CREATIONISTS

Creation advocates are not the only ones called upon to explain the *whys*. Any viable scientific model, whether a creation model or an evolutionary model, attempting to explain life's history must find credible answers to questions such as:

- Why does the structure of the universe, including its physical laws and constants, appear to be planned billions of years in advance for the arrival and benefit of the human species?
- Why are Earth's crust and ocean, and the elements they contain, optimal for advanced life?
- Why are so-called transitional life-forms most abundant among species with the lowest probability to survive mutational and environmental changes and least abundant among species with the highest probabilities to survive such changes?
- Why does life's timing, quantity, type, and diversity throughout the past 3.8 billion years consistently anticipate the needs of future species, including humans?
- Why does life's quantity, kind, and diversity always precisely compensate for changes in the sun's luminosity?
- Why do the laws of physics tend to restrain the expression of evil?
- Why are the exquisitely fine-tuned characteristics of the universe that make a home for humanity possible identical to the exquisitely fine-tuned cosmic characteristics that allow humans to observe the universe's origin and development?
- Why do so many plants and animals exhibit altruistic behavior?
- Why are there so many examples in nature of the sudden appearance of multiple-partner symbiosis?
- Why do humans everywhere distinguish between right and wrong, good and evil?
- Why do humans alone, among all species on Earth, search for a sense of hope, purpose, and destiny?

TESTS OF A MODEL'S STRENGTHS

The process of attempting to develop credible answers to the challenging *why* questions raised by various creation/evolution models provides new opportunities to craft more complete and detailed models. It also reveals new ways to put competing models to the test and yields new predictions based on answers to the *why* questions. This enterprise of developing a second tier of tests for creation/evolution models through predictions will be taken up in the next two chapters.

CHAPTER 10

THE POWER OF PROOF

I kept translating the unbelievable parts into something I could believe until finally I reached the breaking point—I mean my mind just couldn't twist things any more, and somehow I knew it couldn't be true literally, so I just stopped believing and knew it must be a play."[1] This *War of the Worlds* listener likely expressed what countless others experienced. His comments describe the process used to weigh evidence and filter out the implausible, an especially challenging task in evaluating long-held beliefs. The process can be torturous—too much so, in some cases. Sometimes, no matter how much proof a person sees, it's not enough.

At a meeting of Atheists United, attendees were asked, "How many of you would exchange atheism for theism if you could see, over the course of time, a steady increase in the scientific evidence for God?" Roughly two-thirds of those present (a crowd of 300) indicated that they would. In response to the question, "For how many of you would *no* amount of scientific evidence for God ever move you to abandon atheism?" about one-third raised their hands.[2]

An assembly of young-earth creationists was asked, "How many of you would accept that the Earth is several billion years old if you could see increasing evidence to that effect from both science and the Bible?" Only about half said they would. The rest indicated that no amount of evidence could ever change their minds.[3]

When pressed for their reasoning, the no-amount-of-evidence groups said they did not consider abundant evidence adequate. They demanded absolute proof. The irony of this demand lies in the fact that for humans, such proof does not exist. Testing and discoveries can go only so far in bringing resolution to certain debates, including the creation/evolution conflicts. The demand for absolute proof cannot possibly be fulfilled because no human is *all*-knowing about *any* subject. The human capacity to make detections and measurements is confined to the universe's space-time surface. Furthermore, only a portion of that surface has been studied and even that portion has not and cannot be plumbed for all its data. These limitations of human investigation guarantee that absolute proof for any creation/evolution issue never can be attained. No amount of testing can ever deliver it.

PRACTICAL PROOF

Absence of absolute proof, however, is not absence of proof. All humans, including those skeptics demanding absolute proof, draw numerous conclusions and make countless decisions for which they claim certainty based on practical proof alone. Practical proof emerges when the probability for an outcome is determined to be so extremely high that one's confidence in that outcome is for all practical purposes equivalent to absolute proof.

The reliability of the second law of thermodynamics serves as a good scientific example. According to this law, heat flows from hot bodies to cold bodies. And yet statistical mechanics reveals a very tiny probability that heat flow could reverse within a small region of a much larger volume. So, a man coming out of the cold into a warm room faces an extremely tiny possibility of getting colder rather than warmer. But, because that probability is less than one in 10^{80}, that man is justified in ignoring that possibility. It won't happen. In fact, physicists are correct to conclude that such an outcome has not happened and cannot happen anywhere or anytime in the universe, given the limits of its size and age.

PRACTICAL PROOF FOR ORIGINS

An example of practical proof for the creation/evolution conflict is the impossibly low probability (apart from divine intervention) that a body will exist anywhere in the observable universe with the capacity to support physical intelligent life. As noted in chapter 6, 40 years ago astronomers Frank Drake and Carl Sagan used probability theory to support naturalism. They convinced many scientists that the probability may be so high as to render a Creator unnecessary for explaining human existence or, for that matter, the existence of millions of intelligent life-forms on millions of other planets in the Milky Way Galaxy.[4]

Since the 1960s and 1970s, astronomers have determined that life, especially intelligent life, demands dozens of specialized features of its planet, not to mention of its planetary system, star, galaxy, and galaxy cluster. And each of these features must be extraordinarily fine-tuned for life to survive for any length of time—even if somehow placed there. Therefore, Drake and Sagan's probability needed to be recalculated. Recalculations began in 1988. To restate the original question, what is the probability that any one body in the observable universe would by natural means alone possess all the characteristics necessary for the survival of any conceivable physical intelligent life species? The first results were published in 1995.[5] Subsequent calculations were published in 2001 and 2002.[6] (See table 10.1, below, for results.) Ongoing updates are posted at designevidences.org.[7]

Published calculations and updates over the course of 10 years provide a compelling test for biblical creation models. If divine creation did not occur, new scientific discoveries and understandings will cause the list of known characteristics necessary for life's support to shrink. At the same time, the degree of fine-tuning in the remaining characteristics should become less impressive and less significant. Practical proof for the naturalists' story of the universe and life on Earth would thus gain strength.

However, if the Bible's creation story better corresponds to reality, then as scientists learn more about the universe and Earth's life, the number of physical characteristics demonstrating fine-tuning will increase, and the degree of precision life requires of those characteristics will also increase. Practical, tangible proof would thus grow stronger. As table 10.1 illustrates, the trend line thus far strongly favors a biblical creation model.

TABLE 10.1:
INCREASING EVIDENCE FOR FINE-TUNING OF EARTH AND ITS COSMIC ENVIRONMENT FOR INTELLIGENT LIFE*

DATE	NUMBER OF FINE-TUNED CHARACTERISTICS	NATURALISTIC PROBABILITY FOR A PLANET OR MOON THAT CAN SUPPORT INTELLIGENT LIFE
1995	41	10^{-31}
2000	128	10^{-144}
2001	153	10^{-172}
2002	202	10^{-217}
2004	322	10^{-282}

*These probabilities presume that the observable universe contains 10 billion trillion planets and a trillion trillion moons (the maximum numbers that any astronomer has ever proposed). They also take into account that many of the fine-tuned characteristics are dependent rather than independent factors. For a description of the various fine-tuned characteristics, an explanation of how the probabilities were calculated, and citations to the science research papers documenting the relevant discoveries, see the RTB Web site[8] and the RTB books *The Creator and the Cosmos*[9] and *Lights in the Sky and Little Green Men*.[10]

KEY QUESTIONS AS PRACTICAL TESTS

The reliability of proofs for any given model can be augmented (or diminished) by answers to the following questions:

- How quickly and extensively does new research yield further evidence for (or against) the model?
- Are the adjustments to the model in response to new research findings becoming progressively smaller (or greater)?
- From how many disciplines and subdisciplines does evidence for the model arise?
- How many independent means of measurement in a particular discipline produce evidence for the model?
- Are the various measurements of evidence for the model providing increasingly consistent (or inconsistent) results as the measuring accuracy improves?
- Is the model becoming more detailed and comprehensive in its explanatory power as evidence accumulates?
- Does the accumulating evidence for the model yield more research questions/problems for researchers to explore?
- Does accumulating evidence eliminate some competing models and/or compel alteration of other models into forms more closely resembling the model under investigation?
- As research progresses, to what degree are the model's predictions (both short- and long-range) successful?

Vigorous, honest competition among different models mightily encourages and significantly benefits from these questions, as the cosmology example shows in chapter 3 (see pages 44–51).

In the creation/evolution debates, no shortage of competition exists. What's lacking is active, informed, friendly dialogue on these questions and—until now—sufficiently detailed creation models that uphold and embrace the findings of science.

TESTING THROUGH PREDICTIONS

The key to successful application of the nine test questions is the development of well-crafted predictions of what scientists should (in the scientific sense of the word) discover in future research. For predictions to be of any use in bringing resolutions to creation/evolution issues and debates, they must be detailed, distinctive, and comprehensive. The predictions ought not be so vague that the proponent of a particular model runs no risk of being wrong.

A prediction, for example, that many species of life will continue to exhibit a capacity to adapt to changes in their respective habitats is not useful. An advocate for a particular creation/evolution model must predict the limits of adaptation for a particular species in response to specific changes taking place in the species' habitats. Predictions also must estimate how quickly the adaptations will occur, what specific consequences will likely befall the species as it adapts, and, most importantly, how and why the proposed change mechanisms will produce these changes.

Designing predictions to show a difference with respect to competing models permits comparisons. Predictions unique to one model and contrary to all other models hold the greatest promise for bringing resolution on specific creation/evolution issues. Such distinctiveness can quickly reveal whether or not a model should be retained for further development or rejected. The pursuit of distinctive features in models also encourages the necessary creativity to deal with currently unanswered questions.

Finally, predictions must be comprehensive enough to address all (or nearly all) of the major, relevant creation/evolution issues. While no model can hope to explain everything (because human knowledge always will remain finite), a good creation/evolution model needs to provide explanations for already-observed phenomena relevant to the creation/evolution debates. As such, it should produce predictions about what researchers will discover as they continue to study the broad array of creation/evolution disciplines.

In the next chapter a sampling of predictions that arise from the RTB creation model is compared side-by-side with predictions forthcoming from a few of the more prominent competing creation/evolution models. (More exhaustive comparisons have been made or will be made in other RTB books.) As much as possible, short-range predictions are compared so that readers can know within months, or at most a few years, which of the models prove successful and helpful in understanding the truth and which do not. Such successes and failures, while not producing the final complete answer, nonetheless could lead the way in bringing a large measure of understanding and resolution to what has been, for the past 200 years, a most contentious subject.

CHAPTER 11

TESTING CREATION/EVOLUTION MODELS WITH PREDICTIONS

Testing beliefs about creation and evolution definitely has risks, as did testing the reality of a hostile alien invasion. Almost without fail when RTB scientists present their testable creation model in front of an academic audience, some kindly agnostic or atheist scientist will warn of the dangers of putting "faith" to the test. Typical questions and cautions include these:

"Will you really throw away your faith in a Creator if your predictions prove to be incorrect?"

"I can tell your life is better because of your belief in God. I would hate to see that good part of you destroyed by advances in scientific knowledge."

"Wouldn't people of different faiths from yours be spiritually crushed if they discovered their beliefs fail scientific testing? Do you really want to be responsible for that kind of emotional damage?"

"Aren't we really better off just letting people believe whatever they want about God?"

A man who authored 13 books of the Bible answered these questions nearly 2,000 years ago:

If Christ has not been raised, our preaching is useless and so is your faith. More than that, we are then found to be false witnesses about God.[1]

Paul's point is that while people indeed may become emotionally attached to their beliefs, such beliefs are worthless if they are not founded on fact. He adds that it is morally offensive to promote unfactual beliefs. With these points in mind, all the participants in the creation/evolution conflict can put their beliefs to the test with nothing of immeasurable value to lose and everything of enduring value to gain.

Putting My Faith to the Test

Audiences express astonishment when I say I would let go of my Christian faith if it were demonstrated beyond any reasonable doubt that Christianity lacked a factual foundation. That's not to say my life of faith — which began after two years of study and testing — has brought me anything less than great joy and fulfillment and hope. But if these feelings I've experienced are based on delusion and untruths, they are worse than worthless to me. I want to invest my life in reality, whatever it may be, not in an illusion of reality. I would also expect that if my faith is indeed rooted in reality, the factual evidence for my faith will continue to increase as we humans continue to learn more about the universe and humanity's place in it.

CONTRASTING PREDICTIONS

Crafting predictions of what scientists should discover in the future that can differentiate the competing creation/evolution models is no small or simple task. But it must be done for useful evaluations to be possible.

To analyze several creation/evolution models in anything more than a cursory manner would fill many volumes. Given the limitations of this book, only the three most familiar Western models will be compared and contrasted with RTB's creation model:

- Naturalistic evolution models
- Young-earth creationism models
- Theistic evolution models

For definitions and brief descriptions of these models, see chapters 1 and 2. A fourth well-known model, directed panspermia (the hypothesis that extraterrestrial intelligent aliens planted various types of life on Earth at various times throughout the past 3.8 billion years), is addressed in two previously published books.[2]

The comparisons that follow are admittedly superficial. Any inaccuracies in statements of specific predictions emanating from the various models may result from their current lack of detail, scope, and definition. The intent of this work is not to be thorough and complete, but rather to start the comparison-and-contrast process. Future scholarship in developing a variety of detailed models with broader explanatory power will permit more complete and probative comparisons. As various models continue to develop and improve, the task of performing comparative analysis will become more rigorous and conclusive.

The comparisons that follow are also unavoidably dated. The RTB creation model, like other models, is not static. Scholars constantly make small adjustments, additions, extensions, and improvements to their models as new research findings and discoveries accumulate. Such a process of improvement and maturation is the hallmark of any creation/evolution model worthy of consideration. The models compared here are represented as accurately as possible at the date of this writing.

Again, for the sake of brevity, only two sets of predictions for each of the simple and complex sciences are compared and contrasted in any kind of detail. A sampling of other predictions, stated with extreme brevity, appears in table format in appendix F (see pages 227–252). This selection is not meant to imply that creation/evolution issues are strictly scientific. Some predictions emerge from the social sciences, philosophy, and theology as well (see appendix F, pages 248–252).

Predictions List: Simple Sciences

Big Bang Creation Event

The RTB creation model predicts that disputes over the validity of the big bang will diminish as astronomers learn more about the origin and structure of the universe. The model anticipates that new evidences for a big bang creation event will emerge. That is, astronomers should find additional confirmation that the universe traces back to a beginning in finite time, that the universe began in an infinitely or near infinitely hot, dense state, that the universe has been continuously expanding since its beginning, and that the universe has been continuously cooling. What's more, new discoveries by astronomers will increasingly establish that the physics of the big bang creation event accounts for many, though not all, the design features of the universe that make life, human life in particular, possible.

Nontheistic naturalist models of the universe predict that new astronomical discoveries will increasingly establish that the universe does not have an actual beginning but that the universe in some respect will prove to be eternal and self-caused. Such models also predict that new astronomical discoveries will increasingly shake apart the current astronomical consensus that the physics of the big bang event must be exquisitely fine-tuned for life to be possible in the universe.

Young-earth creationist models predict that new astronomical discoveries will prove fatal for all big bang models. Specifically, they predict that evidences for the big bang will be progressively overturned and will lead to more and more astronomers' abandonment of the big bang as a plausible explanation for the origin and structure of the universe. Young-earth models also predict that emerging evidence will show that the cosmic age indicators widely accepted today are illusory and that the universe is only some thousands rather than billions of years old.

Theistic evolution models agree with the RTB creation model in predicting that evidences for the big bang creation model will become more numerous, more compelling, and more consistent. They differ from the RTB creation model in one important respect. They predict that future astronomical discoveries will increasingly demonstrate how the physics of the big bang creation event explain *all* the design features of the universe that make life possible.

Humanity's Special Location for Cosmic Observation
The RTB creation model predicts an increase in astronomical evidence that Earth resides at the ideal location in the cosmos not only for harboring human civilization and technology but also for viewing the totality of cosmic history. This position, according to RTB's model, will increasingly prove rare and ideal for observing the entire history of the universe back to the cosmic origin event itself and for discovering and measuring a multitude of cosmic and galactic design features that make human life uniquely possible on Earth.

Nontheistic naturalist models for the universe predict that new astronomical discoveries will show how unremarkable is Earth's location in the universe both for habitability and for observation. Specifically, the naturalistic models predict that astronomers soon will discover other planetary systems in the Milky Way Galaxy where advanced-life-supportable planets could definitely exist and where observers could view the universe just as easily and as thoroughly as astronomers do from Earth. Likewise, these models predict that astronomers will soon identify many other galaxies where advanced life could exist, and in a position from which the universe would be readily observable.

The predominant young-earth creationist models predict that future astronomical discoveries will increasingly establish galactocentrism, the idea that the Milky Way Galaxy is located at the actual geographical center of the universe.[3] This belief contrasts with RTB's creation model as well as with the nontheistic and theistic evolutionary cosmic models, all of which predict that astronomers in their research will increasingly confirm that the Milky Way Galaxy, like all galaxies and stars in the universe, resides on the surface of the universe much as cities reside on the surface of Earth.

Theistic evolution models agree with the RTB creation model in predicting that astronomers and physicists will continue to accumulate evidence of Earth's unique, ideal location for supporting human civilization and for allowing observation of the universe. However, they differ from the RTB creation model in predicting that future astronomical discoveries will increasingly establish that the big bang creation event and its physics totally account for Earth's favored position for both habitability and cosmic observation.

PREDICTIONS LIST: COMPLEX SCIENCES

Fine-Tuning of Plate Tectonics

The RTB creation model predicts that as scientists continue to research the causes and effects of plate tectonics, their findings will reveal

- more and clearer evidence for the exquisite fine-tuning required by long-lasting, essential-for-life plate-tectonic activity, and
- more and clearer indications of the extreme rarity of plate tectonics anywhere in the universe.

The model also predicts that scientists will find an increasing number of ways in which plate tectonics, at just-right levels, contribute to the support of advanced life and global, high-tech civilization in particular.

With respect to the Flood described in Genesis 6–9, the RTB creation model holds that this event destroyed all mankind but was local in its extent (in that humans were still localized) and that the level of plate-tectonic activity has remained relatively constant over the past several million years. The RTB creation model (as well as the nontheistic and theistic evolutionary models) predicts that research will increasingly confirm the fixity of the laws of physics and of the space-time dimensions of the universe.

Nontheistic naturalist models predict that the evidence for fine-tuned, long-lasting plate tectonics will weaken as scientists learn more about plate-tectonic phenomena. In contrast to the RTB creation model, they predict that planets (or other bodies) with long-lasting plate-tectonic phenomena will prove relatively common in the universe. They further predict that these phenomena will prove less and less crucial to the needs of advanced life and especially of human civilization, and their apparent fine-tuning will eventually be seen as vastly overrated.

Young-earth creationist models predict that all plate-tectonic activity occurred during the past 10,000 years. They expect advancing research to show that the bulk of Earth's tectonic activity took place during the 13 months of a global flood event (roughly 5,000 to 7,000 years ago) and possibly also at the time of Adam's initial rebellion against God (roughly 6,000 to 10,000 years ago).

Young-earth creationist models also expect future scientific research to show that many of the laws of physics and/or the space-time dimensions of the universe were dramatically altered during the Genesis Flood and/or at the time of Adam's initial rebellion against God's authority. These alterations would account, they say, for the dramatic changes in plate-tectonic activity at those events.

Theistic evolutionary models anticipate the increasing ability to prove no interventionist miracles (see chapter 4, pages 71–72) are necessary to explain Earth's plate-tectonic activity or how strategically designed it is for the support of a global, intelligent, high-tech civilization.

Relationship of Hominid Species to Humanity

The RTB creation model predicts that future anthropological and genetic research will increasingly confirm that the human species is biologically distinct and not descended from any of the hominid species. The model predicts a strengthening of evidence for the genetic, anatomical, and behavioral uniqueness of humanity—a set of characteristics that could not have evolved by natural process from any of the hominid or other primate species. Further, the RTB creation model predicts the discovery of additional indications that at least some of the interventionist miracles are necessary to explain humanity's existence and uniqueness.

Nontheistic naturalist models for humanity's relationship to the hominids predicts exactly the opposite of the RTB creation model. They anticipate increasing evidence that humanity's apparently unique attributes and behaviors came by natural descent from one or more of the hominid species. These models further predict that as anthropological and genetic research advances, humanity will prove less and less distinct from the most recent hominid species. Likewise, they predict that evidence for interventionist miracles to explain humanity's unique characteristics will steadily decline.

Young-earth creationist models predict that advancing research will increasingly prove present-day humans anatomically, genetically, and behaviorally identical to Neanderthals, *Homo sapiens idaltu*, and *Homo erectus*. They also expect future discoveries to show that humans and all hominid species have existed within the past 10,000 years (contrasted with RTB and naturalist/evolutionary models, which place humans on Earth about 50,000 years ago and the earliest hominids as many as 6.5 million years ago). Young-earth creationist models agree with the RTB creation model's predictions that increasing evidence will affirm the necessity of at least some interventionist miracles to explain the origin of humanity and the hominids.

Most theistic evolutionary models expect future research to establish that all of humanity's anatomical, genetic, and behavioral characteristics can be explained by divinely directed descent from previously existing species through natural means. Consequently, these models predict that humanity will prove less distinct from the recent hominids than does the RTB creation model, but more distinct from recent hominids than do the nontheistic naturalist or young-earth creationist models.

BRINGING THE CAMPS CLOSER TOGETHER

The predictions outlined in this chapter and throughout this book (including appendix F) will be tested by advancing research, most within the next few years. The relatively short range of these predictions gives them substantial appeal. This testing of predictions holds potential for objective verification/falsification of the major creation/evolution models within a relatively short time. While emotion is always involved with creation/evolution ideas, this approach can help separate fact from fantasy, reduce rancor, and curtail the acrimony of evolution/creation dialogue. Rather than engage in caustic battles, all sides would do well to wait and watch a few months or years to see which models' predictions prove accurate and which do not.

Allowing predictive success and explanatory power to settle creation/evolution differences shifts the emphasis of the discussion from defending the infallibility and finality of one's interpretation to discovering what's true and deepening one's understanding of the truths already uncovered. While to let go of the goal of winning may be difficult, one can only hope that participants in the creation/evolution debates recognize the pursuit of truth as a valuable endeavor—worth the struggle and humility (or humiliation) it requires.

The new goal ideally should be the improvement and extension of models that have survived the rigors of predictive testing. Cutting-edge discoveries and measurements, which bring deeper, more comprehensive understandings of the natural realm, can be used to improve the breadth and depth of surviving models' explanatory power and capacity to generate even more predictions. In other words, the focus stays fixed on learning, improving, and extending the most successful models. Delight in the scientific enterprise is greatly enhanced.

THE NEXT STEPS

One of the most important steps toward resolving creation/evolution conflicts is testing various competing models through specific predictions of what researchers can be expected to discover as they continue to probe the frontiers of knowledge. Even while that testing is in progress, other steps can be taken to remove obstacles and encourage that pursuit. The next chapter points the way.

CHAPTER 12

SPEEDING THE TRUTH QUEST

I t is fear, ultimately, that drives H. G. Wells's story," wrote Chicago journalist John Calloway. And it is fear, ultimately, that fuels much of the evolution/creation tension and hostility. Most of this book has focused on the development and presentation of a testable, verifiable/falsifiable, and predictive creation model, one that invites competition from other similarly framed models. While some attention has been paid to barriers that hinder the development and refinement of such models, a few remaining barriers, including fear, must still be addressed. Moving past them is vitally important to the future of science.

CHRISTOPHOBIA

For several decades, fear of religious radicalism and the encroachment of church into the affairs of state has led to public avoidance of anything that could possibly be construed as an unconstitutional "establishment of religion." Religious pluralism rules the day—with one obvious exception. Christianity seems particularly taboo. Christophobia[1] appears especially pronounced in academia. Special courses and programs on Islam, Hinduism, Native American religions, and other religions, old and new, abound on campuses across North America. Alternatives to a Christian perspective are encouraged while, subtly or overtly, endorsement of biblical views meets with disdain or is flatly forbidden.

In the halls of science, a familiar refrain echoes: Christianity stultifies scientific advance and promotes ignorance, always appealing to the "God-of-the-gaps." The idea that biblical commentary on phenomena within the natural realm may have scientific merit seems utterly ludicrous to many, or most, science scholars and educators (see pages 36–39).

And yet, worldview applied to the scientific enterprise has enormous consequences for the discovery process. A researcher's worldview not only impacts how he or she does science but also what avenues of research he or she deems worth pursuing. Thus, a worldview holds potential to move research forward, to bring investigative progress to a halt, or to cause inefficiencies that significantly slow progress toward discovery.

The assumption that naturalistic evolution governs the history of life on Earth, for example, led to the deduction that the genomes for advanced species predominantly contain useless junk — the accumulation of millions of generations' worth of genetic accidents (see chapter 9, pages 168–169). This inference led to the 30-year abandonment of research into possible functions of non-protein-coding DNA, the so-called junk DNA.

This delay in the study of junk DNA serves as a classic illustration of the "no-God-of-the-gaps" philosophy so pervasive in science. But a gap in human recognition of possible purposes for non-protein-coding DNA collided with the assumption that no creative intentionality imbued it with purpose, and valuable time was lost.

Sometimes, fear of a Christian interpretation of science leads not only to wasted time but also to wasted research talent and money. Astrobiology and SETI (search for extraterrestrial intelligence) serve as examples. In astrobiology, the no-Creator assumption combines with the awareness that life arose on Earth in a geological instant without benefit of prebiotics to compel the conclusion that the origin of life must have been — and still must be — an extremely simple naturalistic event.[2] Based on this conclusion, a new branch of science was born. Astrobiology springs from the conviction that literally millions of planets in the Milky Way Galaxy, as well as several solar system bodies, must be teeming with life.

SETI combines the no-Creator assumption with the observation that technologically advanced humans arose from the first bacteria in only a few billion years. The deduction then follows that technologically advanced civilizations must exist on hundreds, if not thousands, of planets in the Milky Way Galaxy. Faith as fervent as religious zeal has lured huge amounts of public and private funding, not to mention huge chunks of valuable telescope time, into efforts to capture signals from these distant, hoped-for civilizations. For a tiny fraction of the cost and telescope time, the question of whether or not life-supportable bodies are common in the universe and whether or not the molecular building blocks of life

(Photo courtesy of Dr. Seth Shostak/Photo Researchers, Inc.)

(Photo courtesy of Dr. Seth Shostak/Photo Researchers, Inc.)

Figure 12.1:
Part of the Allen Telescope Array (Top)
and the Arecibo Observatory (Bottom)
These superb radio telescopes are often devoted to SETI (the search for extraterrestrial intelligence) rather than to studying the physics of the universe.

(homochiral amino acids and pentose sugars, nitrogenous bases, proteins, DNA, and RNA—see pages 117–123) are ubiquitous in space could be put to the test.

So far, 30 years of intense research investment has yet to yield a shred of evidence for the existence of *any* indigenous extraterrestrial life-building molecules (see pages 117–123) or identify even one extraterrestrial body capable of supporting life (see pages 110–117). Until NASA can demonstrate that extraterrestrial molecular building blocks of life exist in any significant concentrations and that mechanisms exist for their safe transport through interstellar space, it seems a waste of valuable space mission resources to look for indigenous life on Mars or inside Europa. Far more productive, for example, toward understanding life's origin and history would be missions to the moon to recover tiny fossilized remains of Earth's first life (fossils of earliest life that remained on Earth were destroyed by metamorphic processes). These would have been wafted upward, like dust from a beaten rug, whenever Earth suffered a major meteor (or other) impact.[3]

Efficiency and productivity of scientific research can be powerful tools for testing specific creation/evolution models in particular and worldviews in general. If a given perspective delivers scientific discoveries and scientific understanding at a faster pace and for less research investment than a competing worldview, then that perspective can be considered a more likely description of reality. Similarly, if one creation/evolution model generates more scientific breakthroughs, better explanations of natural phenomena, and more comprehensive integration of scientific disciplines for less effort and expense than a competing model, then the former should be awarded appropriate consideration, whatever its philosophical or religious implications may be.

SCIENCEPHOBIA

The fear of science and the avoidance of anything that may encourage trust in conclusions from the scientific community—sciencephobia—has infected many churches and other Christian institutions. Anxiety and distrust often bar any teaching of so-called secular science in these places. In some instances, scientists and science students are banned from participating. Sadly, some churches and schools teach that science is the great enemy of the Christian faith; so students are strongly encouraged to avoid careers in science, even courses in science, lest their faith be undermined.

Thus, sciencephobia prevents influential segments of the Christian community from interacting with or engaging in dialogue with the scientific community. By refusing to acknowledge the important contribution of scientists to the discussion of life's origin and history, these Christians often provoke scientists into aggressive verbal attacks and thus miss or seriously diminish their openness to any constructive criticism of their creation/ evolution views. Such attacks reinforce the presuppositions of sciencephobics that scientists and the scientific enterprise are fundamentally hostile or evil. Such reinforcement leads to even greater polarization.

CENSORSHIP

The scientific enterprise is characterized by objectivity and freedom from censorship. Science research in most disciplines fosters a free-market exchange of ideas and research that fuels innovation and creativity—the hallmarks of the scientific age, past and present.

Bitter attacks and counterattacks unleashed by creation/evolution debaters, though, are threatening this free-market vitality. The verbal assaults represent thinly veiled attempts to prevent those with opposing models or perspectives from being heard. Such attempts at censorship and marginalization often take the form of hard lobbying for a change in the definition of science and/or in the methods of scientific investigation. Ironically, such changes if enacted would undermine the very foundations of the scientific cause. And stranger still, the leading lobbyists are among the nation's most influential educators.

Evolutionist Censorship

Eugenie Scott, executive director of the National Center for Science Education, defines "science" as "an attempt to explain the natural world in terms of *natural* processes, not supernatural ones"[4] (emphasis in original). Lawrence Krauss, director of the Center for Education and Research in Cosmology and Astrophysics at Case Western Reserve University and a prominent spokesman in the fight against the Intelligent Design movement (IDM), echoes Scott's definition: "Science assumes that natural phenomena have natural causes." He concludes, "The concept of 'intelligent design' is not introduced into science classes because it is not a scientific concept."[5]

Scott, Krauss, and most of the science education community see no room for consideration of a supernatural Entity's involvement in the natural realm—no possibility for empirical evidence to be construed as support for any kind of creation model. Some soften their stance by saying it's not that science denies God's existence or His possible role as a Creator, it's just that science is incapable of ever detecting God's operation as a Creator. In Scott's words, "No one yet has invented a theometer; so we will just have to muddle along with material explanations."[6] We cannot test for God, she adds, because we cannot "hold constant the actions of supernatural forces."[7] In other words, science and scientific testing must be limited to direct observations of events occurring in nature or under controlled, laboratory conditions.

Such a definition of science is so narrow, however, that it guts much (if not most) of the scientific endeavor, from theoretical physics to astronomy to paleontology to geophysics to theoretical chemistry to physical anthropology and beyond.

Creationist Censorship

Many creationists play the same science definition shell game, disregarding unwanted findings as "out of bounds." Young-earth proponents say that because no human was present when God created the universe, no scientist has any factual basis for theories of the universe's origin and subsequent development. By this verbal flourish, they discount astronomical observations and relegate physical theories about the big bang, the universe's age, and the development of galaxies, stars, and planets to the realm of pure speculation.

According to Ken Ham, of Answers in Genesis, "Scientists only have the present—they do not have the past."[8] Like Scott, Ham invalidates virtually all the theoretical and observational sciences. On this basis, he and his creationist colleagues claim no need to consider "scientific" evidences that challenge their creationist views.

Some intelligent design advocates play their own version of the shell game. Instead of lobbying for a change in the definition of science, they use a kind of double-speak with respect to their quest. For example, Robert Latimer, a chemist and founder of Science Education for All Ohioans, told a newspaper reporter, "Intelligent design says nothing about religion or about the designer."[9] His group is among those who claim no connection between design theory and religious beliefs. Thus, they quell any discussion of the designer's identity and of how design may have been implemented. In other words, they tend to shut down any attempt to build a testable model for intelligent design.

Open Discussion in the Marketplace of Ideas

Censorship renders the scientific enterprise inefficient at best and ineffective at worst. The only way to keep all fields of research wide open to every thrilling discovery relevant to the origin and history of the universe, Earth, life, and humanity is to hold on to (or return to) consistent, rigorous application of the scientific method. All competing models must be tested by that methodology. No exceptions.

OVER-SPECIALIZATION

The often-lamented over-specialization in the sciences (see pages 58–60) represents a serious obstacle to resolution of evolution/creation conflicts. Some researchers have little if any exposure to those eras of cosmic or terrestrial history in which creation proponents claim divine activity may have occurred. As a result, these scientists lack the opportunity—and the impetus—to put creation/evolution models to the test. They likely see no need to do so.

Even for researchers whose studies do focus on eras of possible divine activity, specialization can pose a problem. One researcher may see evidence that contradicts a particular creation/evolution model in his or her own discipline and yet presume no such contradiction exists in other disciplines. That "glitch" is then treated as an anomaly that future research will likely resolve rather than as a contribution to an accumulation of evidence that could possibly revise or even overthrow that model.

Two examples illustrate this point. In the 1950s and 1960s, origin-of-life research was presumed to be a strictly chemical discipline. While several chemical roadblocks stymied attempts to simulate natural pathways to self-assembly of life molecules, confidence remained high that such pathways would eventually be found. Meanwhile, scientists in other disciplines were demonstrating that the hoped-for chemical successes were essentially irrelevant due to a myriad of geophysical, atmospheric, and other factors. Soon thereafter, origin-of-life research expanded into nearly all the scientific disciplines, where naturalistic scenarios have encountered a host of irreconcilable problems.[10]

At the same time, the question of Earth's age—thousands of years versus billions of years—was thought by many Christians to involve only the discipline of geology. While some serious geological challenges to a young-earth interpretation were acknowledged, young-earth creationists felt secure that their models were in good shape on other fronts. Through the past half-century or so, that security has been gravely disturbed by the recognition that virtually every discipline of science—and even theology—poses multiple challenges and contradictions to the young-earth model.[11]

GOOD, BAD, AND UGLY DEBATES

A number of highly publicized debates between atheist scientists and young-earth creationist or intelligent design advocates have shaped the recent profile of the creation/evolution controversy. While they certainly attract large crowds, these debates tend to do more harm than good in the pursuit of truth. It's not that debates are inherently unproductive, but these certainly have been, for a few obvious reasons.

The audiences have generally been laypeople with neither the specialized education nor training to evaluate the arguments presented. Many or most of those who attend come not to learn and decide on a view but rather to cheer for the speaker who champions their view. They come to witness a contest between verbal gladiators. In this atmosphere, of course, the debaters feel compelled to win at all costs. Thus, they refuse to concede any points made by the opponent(s) to their position. Rather than resulting in resolution, such debates tend to increase the level of polarization, hostility, and entrenchment.

Debates, however, can be made productive if they are structured according to the principles of the scientific method. Starting with acknowledgment that no one possesses all the answers, each presenter should be prepared to learn from the others. Because few issues are as simple as an either/or question, most of these debates would do well to involve more than two positions and presenters.

The audience for such debates should be well informed on the issues, having an educational level roughly equivalent to that of the debaters. Attendees should be allowed opportunities for significant interaction with those presenting the arguments. This type of exchange will be most productive if the audience and the debaters place a higher priority on understanding and resolution than they do on "winning." In fact, winning should be defined as progress, of any magnitude, toward resolution of conflicting views.

The length of such debates is also crucial. If time is short, a presenter may be able to raise so many challenges against another position that the defender runs out of clock before he or she can adequately respond. Similarly, a debater may muster up enough points in favor of a position to sound convincing for 20 or 30 minutes—but not for 60 or 90. For these reasons, it's important that debaters not be racing the clock, nor should they be unfairly protected by it. Timekeeping must not take priority over thoughtful evaluation of the evidence.

If seeking some resolution of creation/evolution conflicts is the goal rather than humiliating the opposition, the quality of the audience's knowledge must take priority over its size; and the qualifications of the debaters must take priority over their fame or notoriety. An extended debate in which multiple positions are presented and compared in front of a small, highly trained audience representing multiple disciplines or subdisciplines—that's the kind of debate that can contribute to progress and point the way for future research.

SCIENCE EDUCATION CRISIS

The current science education crisis in North America, in the United States in particular, poses yet another barrier to resolution of the evolution/creation wars. The alarming trend line shows that fewer and fewer Americans are enrolling in science courses at the high school, college, and graduate levels. Therefore, the general public is declining in ability to evaluate critical creation and evolution arguments. But the situation can still be reversed, especially with a change in the approach to evolution/creation issues and questions.

That America now faces a serious science education crisis is beyond question. South Korea, with only one-sixth the population of the U.S., graduates as many engineers.[12] And the situation is considerably worse than this statistic suggests because a large percentage, approaching a majority, of America's science and engineering graduates are from other countries.[13] In a 2003 address to the President's Council of Advisors on Science and Technology, Richard Smalley, Nobel laureate in chemistry (1996), forecast that "by 2010, if current trends continue, more than 90 percent of all scientists and engineers in the world will be living in Asia."[14]

A possible explanation for this sharply declining enrollment in the sciences by Americans and the sharply increasing enrollment by Asians comes from scholars in the People's Republic of China. Chinese paleontologists studying the famous Cambrian explosion fossils in the Chengjiang shale in Yunnan province have commented to their American colleagues, "In China we are not allowed to criticize our government leaders, but we are free to criticize Darwin and Darwinism. In your country you are free to criticize your government leaders, but you are not permitted to criticize Darwin or Darwinism."[15]

Could it be that at least to some degree the science education crisis in America stems from the quelling of controversy, for example, from a dismissive attitude toward any alternative to strict naturalism, so much so that science education has become boring? Is the fear of being religious or philosophical about the implications of the amazing new discoveries at the frontiers of scientific research prompting science educators in America to quell discussion of the really important and intriguing *why* questions? Are science educators avoiding the very issues that could engage their students? (See "Making Science Exciting Again," page 200.)

The potential for "yes" answers to these questions seems reason enough for science educators to include evolution/creation content in their curricula. Students' enthusiasm for science can be rekindled if they realize that science not only can validate or invalidate the details of nature but also can grapple with the really important and meaningful questions of life.

It goes without saying that public instruction on creation/evolution issues must be handled carefully and professionally—with respect for science, the scientific method, and scientific integrity. This simultaneously passionate-about-science and yet objectively dispassionate demeanor may be difficult to inculcate and maintain, but the difficulty must not stop science educators from making the effort.

The result of making science education more engaging and exciting, not just for the "born scientists" but for everyone, is that the American people will become better equipped to evaluate creation/evolution assertions. In addition, they will be better able to personally apply what they learn. As more of the creation/evolution mysteries are solved and as new ones emerge, science education will become even more exciting and thus attractive to students.

Making Science Exciting Again

Fear of controversy, fear of what's new, and fear of getting too religious or philosophical may indeed be making science dull, as the following personal example illustrates.

A few years ago, I was invited to speak at a large aerospace company in Southern California. My subject: "Creation and Evolution: New Scientific Evidence for God." When I arrived, the 400-seat auditorium was packed to overflowing. After my talk, many people stayed for more than an hour to ask questions, despite being required to make up the time.

For the rest of the day, I was told, employees were buzzing about the topic and the points raised in the Q&A session. As I was leaving, one of the company's executives, a self-identified atheist, engaged me in a brief conversation. He told me the company had been sponsoring science enrichment lectures for the employees for a long time. Attendance, he said, usually hovered around 10 to 15, and nobody stayed after the lectures. Looking me straight in the eye, he added, "I may not like this Jesus Christ stuff [I had mentioned Jesus' name only once], but if that's what it takes to get our employees interested in science, I am all for it."

My experience at this setting was not an anomalous occurrence. It has been repeated many times through recent years in schools, high-tech firms, professional meetings, and elsewhere—and not because of any great communication skills on my part.

EDUCATING THE EXPERTS

Laypeople are not the only ones who need to be brought up to speed on creation/evolution issues; scientists, philosophers, and theologians who address the topic also need to be updated. Most scientists have taken few if any courses in philosophy and theology and freely admit to knowing very little about the Bible. Most philosophers and theologians have taken few if any science courses beyond the minimum general education requirement. This isolation explains why science-theology dialogue is so daunting. It also explains why so many scientists, philosophers, and theologians involved in developing creation/evolution models fail to appreciate the degree to which the scientific enterprise overlaps the theological.

Exposing philosophers and theologians and their students to more science education (and scientists) and vice versa, especially at the higher levels, may dramatically lower the level of animosity in creation/evolution debates. Such exposure also can facilitate the kind of cross-disciplinary integration that is key to the pursuit of truth.

ENCOURAGING THE INNOVATORS

A recent editorial in the British journal *Nature* offers what may seem a surprising comment:

> Public controversies that involve scientific uncertainty can be influenced by mavericks. Open confrontation and analysis serves the public better than excommunication.[16]

This observation certainly applies to the creation/evolution debates. Many people already suspect the science establishment of actively suppressing "alternative" theories and of isolating nonconformists. A willingness to go to extraordinary lengths to ensure that new ideas, novel hypotheses, and innovative research obtain fair hearings and appropriate testing (regardless of their source) only makes sense.

An even more significant point is that the issues involved with the origin and history of the cosmos, Earth, life, and humanity are badly in need of new insights. Creativity and innovation may help plow through the many impasses that remain. Again, ensuring that no viable contribution is ignored makes sense for all concerned.

HUMILITY

Perhaps the most critical antidote for polarization and bitterness is humility. When speakers and writers allow for viewpoints different from their own, listeners and readers become willing to consider new evidence and new perspectives and to follow where the combined weight of old and new evidences lead.

Humility in creation/evolution debates involves the recognition that no model has all the answers because no model is — or can be — complete. Consequently, all models are in need of improvement and there is always more to learn. Just as in personal relationships, however, the exercise of humility in creation/evolution debates involves a delicate balance. Giving a model either more or less credit than it deserves, based on its scientific or philosophical appeal, is an expression of false humility

A proper balance in assigning credit where it is due probably will remain one of the biggest challenges in fostering resolution of creation/ evolution conflicts. Simply being aware of the need, however, by itself may facilitate more efficient pathways to truth.

BUILDING BETTER MODELS

To date, serious comparison and evaluation of creation/evolution models has been hampered by the lack of one or more creation models with sufficient detail, breadth, and testable/verifiable/falsifiable scientific content to be judged on the merits of explanatory power and predictive success. For that matter, sufficiently developed evolutionary models are also in short supply. Every evolutionary model currently available can be made much more testable, verifiable/falsifiable, and predictive.

This book places a biblical creation model on the table. In doing so, it invites the development of other models, for both creation and evolution, in forms that are equally open to testing. Given such development, the process of comparison, evaluation, and refinement can yield significant advances toward better understanding of the origin and history of the universe, life, and humanity.

RESOLUTION IN SIGHT

The testable-model strategy as outlined in this book will not end the long-standing creation/evolution conflicts overnight. However, it holds real potential for resolving the most contentious issues within the current creation/evolution controversy in the relatively near future. All other strategies have failed.

Any successful resolutions will indeed raise new creation/evolution questions and puzzles for researchers to explore and study and solve. In this sense, the creation/evolution debates will never end. What can end, though, is the dreadful war of angry words and bitter disputes over the same tired issues. Acrimony can abate as participants see real and steady progress toward answering some of life's biggest questions.

CREATION/EVOLUTION
VERBAL WARFARE

Competition among models—with all the debate, argument, and disagreement such competition brings—invigorates the scientific enterprise and leads to progress in understanding the realm of nature. Typically, researchers view such free-market competition as something that benefits all participants in the pursuit of truth.

However, the subject of creation/evolution has generated competition that is far from collegial. Even the most distinguished leaders and researchers have been caught up in name-calling, insults, and character assaults. The verbal volleys cited below serve to illustrate the acrimony of the creation/evolution debates and the resultant polarization.

CREATIONISTS ON THE ATTACK

Fear of evolution, specifically recognition of the way naturalistic ideology undercuts the value of human life, spurs many creationist leaders to blast the scientific community and evolutionists in particular. Henry Morris, past president of the Institute for Creation Research (ICR) and for many years the leading "young-earth" spokesman, blames evolutionists for "the bitter fruits of widespread amorality, materialism, the drug culture, abortionism, pornography, social diseases and a host of other ills—not to mention communism and fascism."[1] Morris's son John, current ICR president, agrees with this characterization and adds that evolution has produced "not one good fruit in the form of real scientific advance in either living standards or altruistic behavior."[2]

Young-earth creationists openly question the validity of all scientific findings on the origin and history of the universe and life. John MacArthur, pastor and president of The Master's College and Seminary, attacks scientists' capacity to reason. He wrote, "Evolution is as irrational as it is amoral."[3] Jason Lisle, an astronomer with Answers in Genesis-USA, declares, "Many Christians do not realize that the big bang is really just a story about the past that is based on untested and anti-biblical *assumptions*. . . . In a way, the big bang and evolution *are* really science fiction, even though they are taught as fact"[4] (emphasis in original).

INTELLIGENT DESIGN THEORISTS JOIN IN

Frustrated by the dominance of naturalism and the barriers to presenting alternative views on the topic of origins, the Intelligent Design Network in their brochure *Seeking Objectivity in Origins Science* accuses mainstream scientists of trying to "censor the evidence of design."[5] The lack of opportunity to present scientific data that points to intelligent design, especially in the media and education, they say, gives naturalism "a monopoly on the scientific explanation of origins."[6] They charge scientists with abandoning "an objective search for the truth."[7]

The most prominent spokesperson for the Intelligent Design movement (IDM), Phillip Johnson (a University of California, Berkeley law professor), argues that "Darwinists so want the theory to be true they obscure the evidence."[8] In a number of articles published in major newspapers and magazines, Johnson charges them with replacing "real science" with "flim-flam."[9] He wrote, "When our leading scientists have to resort to the sort of distortion that would land a stock promoter in jail, you know they are in trouble."[10]

EVOLUTIONISTS COUNTERATTACK

Fearing that bad science (or antiscience) and religious dogma will creep into the nation's educational system, evolutionists have lashed back. In *Darwinism Defended*, Michael Ruse describes creationism as "a grotesque parody of human thought, and a downright misuse of human intelligence."[11] Stephen Jay Gould, in his book *The Structure of Evolutionary Theory*, refers to creationists as "ignoramuses."[12] Niles Eldredge, coinventor (with Gould) of the punctuated equilibria hypothesis, says, "Creationist literature reads like the worst of the supermarket tabloids."[13] And, in *Telling Lies for God*, Ian Plimer makes this accusation: "Creation 'scientists' engage in blatant scientific fraud" and "fabricate data."[14]

Intelligent Design (ID) advocates have been specifically singled out for attack. Eugenie Scott, executive director of the National Center for Science Education, criticizes the IDM for a "lack of candor in the presentation of the ID case."[15] Physicist Lawrence Lerner complains that "intelligent design creationists often hide the essentially theological nature of their agenda."[16]

Evolutionary biologist Jerry Coyne refers to the research and writing from ID scholars as "stealth creationism," noting that ID proponents "hide their own differences about issues such as the fossil record and the age of the Earth" and "avoid at all costs revealing their own theories about the history of life."[17] Physicist Mark Perakh accuses certain ID leaders of "disdainful dismissal of all and every criticism."[18]

APPENDIX B

DOES THE CONSTITUTION BAR CREATION TEACHING?

The issue of creation versus evolution has been raised in numerous court cases, and there's no apparent end in sight to litigation both in America and abroad.[1] In the famous Scopes Trial of 1925, evolutionists battled to be allowed to teach their view in public schools. Creation advocates fought hard to prevent them from doing so. Initially they succeeded, but soon evolution prevailed and, gradually over time, the right to teach any form of creation or divine design in the classroom evaporated. Or did it?

Through the past several decades, young-earth creationists have sought redress through political action. Many have lobbied their local school boards and state legislatures for "equal access." Their demand is clear: Whenever and wherever evolutionary theory is taught in the public education arena, creationism must be taught at an equivalent content level.

Examining the legal history of the problems and specific court cases brings clarity to the issues involved. A review of the results may help guide future efforts toward resolution of creation/evolution debates and reveal the glaring need to rebuild a sense of teamwork in research.

FIRST AMENDMENT

The First Amendment of the U.S. Constitution has been a critical element in court decisions handed down since *Tennessee v. John T. Scopes*. The First

Amendment simply states:

> Congress shall make no law respecting an establishment of reli-
> gion, or prohibiting the free exercise thereof; or abridging the
> freedom of speech, or of the press; or the right of the people peace-
> ably to assemble, and to petition the Government for a redress of
> grievances.

Philosopher Robert Pennock notes that those opposed to creationism
universally claim that "teaching Creation in the public schools would violate
the First Amendment's establishment clause."[2] Two large nonprofit organizations, Americans United for Separation of Church and State and People for
the American Way, have interpreted this conclusion as a mandate to establish an impenetrable wall of separation between religion and the state.[3]

The mainstream perception that teaching any aspect of creation in
public education is forbidden by the Constitution rests on four court decisions: *Epperson v. Arkansas* (1968), *McClean v. Arkansas* (1981), *Aguillard
v. Treen* (1983), and *Edwards v. Aguillard* (1987). In all four of these cases,
"creation teaching" collided with the First Amendment's establishment
clause—but for reasons that few people fully comprehend.

Epperson v. Arkansas (1968)

Public school teacher Susan Epperson, a 10th-grade biology teacher
at Central High School in Little Rock, Arkansas, brought suit against
an old anti-evolution statute. The law prevented teachers in any state-
supported school or university from teaching or using a textbook teaching
"that mankind ascended or descended from a lower order of animals."[4]
The State Chancery Court found in favor of Epperson based on the free-
speech clauses in both the First and Fourteenth Amendments of the U.S.
Constitution.

However, upon appeal, the State Supreme Court upheld the anti-
evolution statute as within Arkansas' power to specify public education
curriculum. When the case was appealed again, the U.S. Supreme Court
ruled in favor of Epperson. In support of their ruling, the justices noted:

1. Arkansas had no scientific reason, in fact, no reason at all, for
   upholding its anti-evolution statute other than its observation
   that a particular religious group considered evolution theory to
   conflict with its interpretation of man's origin as set forth in the
   opening chapters of Genesis.[5]
2. Arkansas' right to prescribe public education curriculum does not

include the right to prohibit the teaching of scientific data, theory, or doctrine simply because it might offend someone's religious beliefs.[6]

In siding with Epperson, the U.S. Supreme Court sent a strong message that a particular religious viewpoint could not, by itself, be used to hinder the advance of scientific knowledge and understanding.

Although a case can be made that America's Founding Fathers intended for the nation's governmental organization, policies, and laws to be under the authority of the God of the Bible and to reflect Christian principles (see "What Did the U.S. Founding Fathers Intend?" below), they did not encourage or permit public instruction of unscientific teachings.[7]

What Did the U.S. Founding Fathers Intend?

Many people, including such groups as People for the American Way and Americans United for the Separation of Church and State, take Thomas Jefferson's comment about "building a wall of separation between church and state" to imply that the government, government officials, and school-teachers must avoid any public practice, teaching, or support of religious views. However, two facts make this interpretation of Jefferson's words questionable: (1) He originally wrote them to reassure the Baptists in Connecticut that their right to publicly practice and teach their religious views both inside and outside their churches was protected by the U.S. Constitution. (2) Jefferson, in his official capacity as the president of the United States, closed his letter with a prayer to the God of the Bible.[8]

While the Founding Fathers of the United States did not want a federally sponsored church, they opposed "a wall of separation" between God and state — specifically between religion and public life, between Almighty God and political concerns. The Declaration of Independence, itself, makes four direct references to the biblical God.[9]

George Washington, in his capacity as the nation's first president, issued a proclamation of national thanksgiving. He and both houses of Congress recommended to the people of the United States "a day of public thanksgiving and prayer to be observed by acknowledging with grateful hearts the many signal favors of Almighty God."[10]

The U.S. Senate has always had a chaplain paid by the U.S. government. This chaplain has always been a Christian who opens each session of the Senate with prayer.[11] Likewise, the U.S. military, from the earliest days of the Revolutionary War to the present, has provided chaplains (including Christian chaplains).

McClean v. Arkansas (1981)

In 1981, the Arkansas legislature passed a law to mandate that whenever "evolution-science" was taught in public schools, "creation-science" must be taught as well (for definitions, see "Balanced Treatment Act of Arkansas, #590 of 1981," below). This Balanced Treatment Act appealed to the freedom clauses in the U.S. Constitution and its amendments.

Balanced Treatment Act of Arkansas, #590 of 1981

An act to require balanced treatment of creation-science and evolution-science in public schools; to protect academic freedom by providing student choice; to ensure freedom of religious exercise; to guarantee freedom of belief and speech; to prevent establishment of religion; to prohibit religious instruction concerning origins; to bar discrimination on the basis of creationist or evolutionist belief; to provide definitions and clarifications; to declare the legislative purpose and legislative findings of fact; to provide for severability of provisions; to provide for repeal of contrary laws; and to set forth an effective date.[12]

"Creation-science" and "evolution-science" are defined in the Act as follows:[13]

(a) "Creation-science" means the scientific evidences for creation and inferences from those scientific evidences. Creation-science includes the scientific evidences and related inferences that indicate: (1) Sudden creation of the universe, energy, and life from nothing; (2) The insufficiency of mutation and natural selection in bringing about development of all living kinds from a single organism; (3) Changes only within fixed limits of originally created kinds of plants and animals; (4) Separate ancestry for man and apes; (5) Explanation of earth's geology by catastrophism, including the occurrence of a worldwide flood; and (6) A relatively recent inception of the earth and living kinds.

(b) "Evolution-science" means the scientific evidences for evolution and inferences from those scientific evidences. Evolution-science includes the scientific evidences and related inferences that indicate: (1) Emergence by naturalistic processes of the universe from disordered matter and emergence of life from nonlife; (2) The sufficiency of mutation and natural selection in bringing about development of present living kinds from simple earlier kinds; (3) Emergence by mutation and natural selection of present living kinds from simple earlier kinds; (4) Emergence of man from a common ancestor with apes; (5) Explanation of the earth's geology and the evolutionary sequence by uniformitarianism; and (6) An inception several billion years ago of the earth and somewhat later of life.

Concerned about the damage this specifically young-earth creation teaching would do to the credibility of the Christian faith, science education, and the scientific endeavor, the Rev. Bill McClean and a coalition of scientists and science educators filed suit. The Arkansas Court quickly observed that the act mandated a contrived dualism, the dogmatic claim that only two possibilities existed for explaining the history of the universe, Earth, and life on Earth: atheistic materialism and young-earth creationism.[14] Therefore, the court ruled that the exclusion of other possible explanations, by themselves, violated the freedom clauses the act supposedly sought to defend. Ironically, two witnesses (for the defense of the act), Norman Geisler and Chandra Wickramasinghe, presented two different explanations (old-earth creationism and pantheism/deism) for life's origin and history.

The court noted that the freedom clauses in the United States Constitution and its amendments already allow for religious ideas and concepts to be taught in a public education context provided that such ideas and concepts have at least some demonstrable secular merit and academic integrity. If the defendants' young-earth creation science could demonstrate a value independent of its particular religious perspective, then public education access for this specific brand of creation science could not be denied. Consequently, the court focused on the scientific credibility of young-earth creation science.

The scientists who testified for the plaintiff declared (without exception or equivocation) that they could find no merit, credibility, or integrity in the brand of creation science being promoted by the defendants. In response to the question, "Are you aware of any scientific evidence to indicate the earth is no more than ten million years old?" Brent Dalrymple of the U.S. Geological Survey replied, "None whatsoever."[15] When asked, "In your professional opinion, are [sic] the creation scientists' assertions of a young earth been falsified?" Dalrymple stated, "Absolutely, I'd put them in the same category as the flat earth hypothesis and the hypothesis that the sun goes around the earth. I think those are all absurd, completely disproved hypotheses."[16]

In a surprising move, the defense chose not to place on the witness stand either of the two leading young-earth creation scientists of the day, Henry Morris and Duane Gish of the Institute for Creation Research. Apparently the defense did not believe that either Morris or Gish could withstand cross-examination.[17] Instead, the defense called British cosmologist and origin-of-life researcher Chandra Wickramasinghe to testify on behalf of the "creation-science" position.

In a prepared statement, Wickramasinghe explained to the court why he and a number of his colleagues rejected Darwinian evolution as a valid scientific explanation for either life's origin or its history on Earth.[18] However, under cross-examination, Wickramasinghe fully corroborated the testimony of the plaintiff's science expert witnesses that "no rational scientist would believe the earth's geology could be explained by reference to a worldwide flood or that the earth was less than one million years old."[19] Afterward Judge William Overton wrote, "The Court is at a loss to understand why Dr. Wickramasinghe was called in [sic] behalf of the defendants."[20]

Once it had determined that the defendants' young-earth creation science "has no scientific merit or educational value as science,"[21] the court was forced to evaluate the worth of the Balanced Treatment Act on religious merit alone. Unfortunately for the defense, a number of Christian theologians and clergy representing a wide spectrum of denominations, including Presbyterians and Southern Baptists, testified to their disagreement with the defendants' interpretation of the biblical creation accounts.[22] Consequently, the court determined that the act violated the United States Constitution's First Amendment.

In the Court's opinion, the Act would force upon public education students and teachers one particular creation perspective to the exclusion of all others.[23] Therefore, the conclusion that the act violated the First Amendment was inescapable.

Aguillard v. Treen (1983)

In spite of (or perhaps because of) what happened in Arkansas, the state legislature of Louisiana in 1982 passed a Balanced Treatment Law (later dubbed the "Creationism Act") virtually identical to the one overturned in Judge Overton's court. As in Arkansas, a suit was filed within months of the law's passage.

In 1985 U.S. District Judge Adrian Duplantier declared Louisiana's law null and void by summary judgment, that is, without benefit of a trial. He agreed with Overton that the young-earth creation science mandated by that law was not science but rather a particular religious dogma. Duplantier refused to consider any of the thousand-plus pages of scientific documentation (pro and con) filed in the case.

The decision that Louisiana's Balanced Treatment Law violated the U.S. Constitution's First Amendment then went to the U.S. Court of Appeals, where a panel of 15 judges voted (eight to seven) in favor of upholding Duplantier's verdict. Because of the closeness of the vote, the state of Louisiana filed a "jurisdictional statement" with the U.S. Supreme Court, arguing that the Balanced Treatment Law and the response of the lower courts established a substantial federal question. The U.S. Supreme Court justices concurred and agreed to hear the case, which had become *Edwards v. Aguillard*.[24]

Edwards v. Aguillard (1987)

The most famous and illustrative of the young-earth creation science court battles is *Edwards v. Aguillard*, the U.S. Supreme Court's judgment on Louisiana's "Creationism Act." The Supreme Court justices declared the act superfluous in that it did nothing to further its stated purpose, nothing to further the "protecting of academic freedom."[25] More specifically, the justices pointed out that the act failed to "further the goal of 'teaching all of the evidence.'" They added that "requiring the teaching of creation science with evolution does not give schoolteachers a flexibility that they did not already possess to supplant the present science curriculum with the presentation of theories, besides evolution, about the origin of life."[26]

According to the justices, the act would have actually diminished academic freedom. They remarked that "under the Act's requirements, teachers who were once free to teach any and all facets of this [creation/evolution] subject are now unable to do so."[27]

The Supreme Court justices also challenged the fairness concept of equal access. They observed that

> the Act evinces a discriminatory preference for the teaching of creation science and against the teaching of evolution by requiring that curriculum guides be developed and resource services supplied for teaching creationism, but not for teaching evolution, by limiting membership on the resource panel to "creation scientists," and by forbidding school boards to discriminate against anyone who "chooses to be a creation scientist" or to teach creation science, while failing to protect those who choose to teach other theories or who refuse to teach creation science.[28]

Most telling of all, the Supreme Court justices concluded that the "appellants had failed to raise a genuine issue of material fact"[29] or to establish that their viewpoint "constitutes a true scientific theory."[30] Consequently, the court held that "the Act is facially invalid as violative of the Establishment Clause of the First Amendment, because it lacks a clear secular purpose."[31] The panel further pronounced that "while the Court is normally deferential to a State's articulation of a secular purpose, it is required that the statement of such purpose be sincere, and not a sham."[32]

As in the previous state-level court judgments, the Supreme Court justices could find no secular (i.e., scientific) merit in the brand of creation science being defended and so were forced to judge the case on whether or not the act gave unfair advantage to one religious doctrine over another. They ruled that "the Act's primary purpose was to change the public school science curriculum to provide persuasive advantage to a particular religious doctrine."[33] Thus, a majority of U.S. Supreme Court justices (seven to two) found the act in violation of the free exercise of religion clause of the U.S. Constitution's First Amendment.

Faulty Science, Closed Door

In the four most important creation/evolution court battles of the past 50 years, the creation side lost every time. Why? Because its defendants could not prove that young-earth creationism had scientific legitimacy. The Supreme Court justices made their decisions crystal clear. If young-earth creation science is valid as science, its right to be included in the public school curriculum would be assured no matter what its connection to one form of religion or another.[34] Even if young-earth advocates could prove that their position had no connection whatsoever with religion, this theory still would be considered illegitimate due to its lack of scientific integrity. Thus, the appeal to protected access in science coursework has been denied.

According to all four court judgments, scientific merit sets the primary standard. The courts acknowledged that this merit does not mean scientific perfection. If it did, no scientific model or interpretation could be taught. However, creation proponents must demonstrate at least some level of scientific merit. Without it, access to the public education science curriculum cannot be justified.

In the context of scientific credibility, these court judgments against young-earth creation science cannot be construed as the audacious judicial moves many people make them out to be. The courts have been consistent. Since the founding of the United States, and in light of the First Amendment's free-exercise clause, American education has always discriminated against religious ideas and doctrines that lack factual support.

The flat-Earth doctrine provides a good example. Until the 1920s, several religious groups in the United States believed that the Bible teaches a flat Earth.[35] Such religious convictions, however, did not sway the nation's educational institutions to allow flat-Earth physics in their science curricula.

Geocentrism—the belief that the sun, planets, and stars revolve around the earth—serves as a more recent illustration. Several creationist groups assert that the Bible teaches geocentrism and have founded nonprofit organizations to promote this view.[36] Nevertheless, though geocentrism has been part of the American religious landscape throughout the past 250 years, geocentric physics has never been taught as a potentially valid hypothesis in public education textbooks or curricula.

Though each of their decisions has barred the teaching of young-earth creation science, the courts, including the highest court in the land, have given assurances that the door remains open to scientifically credible creation models.

GOOD SCIENCE, OPEN DOOR

At the beginning of the 20th century, the reigning cosmological paradigm posited that the universe was infinitely old, infinitely large, and chemically static and thus that the conditions necessary for operation of life's chemistry remained unchanged over infinite time. Though this cosmic model essentially endorsed one main ideology—atheistic naturalism—to the exclusion of all others (especially the creation doctrines of Christianity, Islam, and Judaism), it generated no First Amendment legal challenges.

No matter where or how far away they looked, astronomers saw a universe that appeared endless and homogeneous. Though scientific evidences seemed compelling, both astronomers and physicists at the time acknowledged that only a tiny piece of the observable universe had been explored. They were also aware that Newtonian mechanics alone, as solidly established as they were, could not explain certain phenomena in the celestial mechanics of the solar system. So they continued their investigations and tested their ideas to see which ones corresponded to reality.

When Einstein's general theory of relativity arrived in 1916 and an optical telescope 100 inches in diameter became available in 1917, scientists made the stunning observation that the universe is not static. Later, even larger telescopes and new technologies revealed that the universe's properties were radically different billions of years ago and that the universe had a beginning in finite time.

During the last 30 years of the 20th century, a number of mathematical theorems based on general relativity confirmed that even space and time had a beginning—coincident with the beginning of the universe.[37] During the past 25 years, astronomical measurements have established the Jewish, Christian, and Islamic doctrines of a universe with a single beginning. (This evidence contradicts the Hindu–Buddhist–New Age doctrine of a universe that cycles through an infinite or near infinite succession of beginnings and endings.)[38]

Regardless of the theological/philosophical implications of modern cosmological research and despite how strongly recent cosmological advances discriminate against the beliefs of certain individuals and major religious bodies, the legislatures and courts of the United States have chosen not to intervene. At least for the hard sciences, legislative and judicial leaders have acknowledged that while the First Amendment cannot be used to protect bad science, neither can it be used to banish good science, whatever its theological implications.

BIBLICAL ORIGINS OF THE
SCIENTIFIC METHOD

A major source of optimism for resolution of the creation/evolution debates, or at least for significant progress toward resolution, is that all the participants in the debates appear to agree on the best method for testing models. That method is popularly termed the scientific method, though a more accurate label would be the biblical method.[1]

The Bible alone among the "scriptures" or "holy books" of the world's religions strongly exhorts readers to objectively test before they believe. According to the apostle Paul, no teaching is to escape testing:

Test everything. Hold on to the good.[2]

Paul exhorts us that such testing, to be effective, will require objectivity, education, and training:

Do not conform any longer to the pattern of this world, but be transformed by the renewing of your mind. Then you will be able to test and approve what God's will is.[3]

Testing before believing pervades both the Old and New Testaments and forms the heart of the biblical concept of faith. The Hebrew word for faith, 'emûna, means a strongly held conviction that something or someone is certainly real, firmly established, constant, and dependable.[4] The Greek word for faith, *pistis*, means a strong and welcome conviction of the truth of anything or anyone to the degree that one places deserved trust and confidence in that thing or person.[5] In every instance, faith in the Bible connotes the response to established truth. Just as there is no faith, from a biblical perspective, without an active response,[6] neither is there faith apart from established truth(s).

Christian scholars throughout church history, from the early church fathers, to Renaissance naturalists, to Reformation theologians, to present-day evangelical scientists, philosophers, and theologians, have noted a pattern in biblical narratives and descriptions of sequential physical events: the Bible authors typically preface such depictions with a statement of the frame of reference (point of view) and initial conditions and then close with a statement of the final conditions and conclusions about what transpired. The Scottish theologian Thomas Torrance has both written and edited book-length discussions of how Christian theology, and Reformed theology in particular, played a critical role in the development of the scientific method and the amazing advances achieved by Western science.[7]

FUNCTIONAL ROLES OF "JUNK" DNA

For more than 30 years, scientists referred to DNA that does not code for the manufacture of proteins as "junk" DNA. This so-called junk is now known to serve several life-critical functions.

Geneticists have discovered five kinds of non-protein-coding DNA: pseudogenes, SINES, LINES, endogenous retroviruses, and LTRs. The term *pseudogenes* comes from the assumption that certain DNA segments are the dead, useless remains of genes that many generations ago did code for proteins. Recent experiments, however, show that many pseudogenes are not useless. When certain pseudogenes are turned off, the organism suffers either fatal or injurious consequences.[1] Geneticists now realize that these pseudogenes somehow protect the protein-coding genes from breakdown or malfunction. Other pseudogenes, they have found, actually encode for functional proteins.[2] Still other pseudogenes were also misidentified and later found by researchers to encode for the construction of molecules at first wrongly considered to serve no purpose.[3]

SINES is an acronym for short interspersed nuclear elements. Emerging research shows that these DNA elements serve at least two distinct purposes. Some help protect the cell when it experiences stress.[4] Others help regulate the expression of the protein-coding genes.[5]

LINES is an acronym for long interspersed nuclear elements. Recent findings show that some LINES play a central role in X-chromosome inactivation.[6] When such inactivation fails, serious genetic disorders result.[7] Another discovered LINES function is to turn off one of the two protein-coding genes inherited from an individual's parents.[8]

Evolutionists once presumed that all endogenous retroviruses were the product of retroviral infections. They hypothesized that retroviral DNA becomes incorporated into the host's genome. New research, however, shows that many endogenous retroviruses protect the organism from retroviral infections by disrupting the life cycle of invading retroviruses.[9] Others function as protein-coding genes.[10]

LTRs, an acronym for long terminal repeats, were once thought to originate from endogenous retroviruses. Recent studies show that several LTRs play crucial roles in protecting organisms from retroviral attacks.[11] Other research demonstrates that some LTRs help regulate the expression of certain protein-coding genes.[12]

Far from being junk, non-protein-coding DNA serves many amazing life-beneficial purposes. These purposes would never have been discovered and understood if geneticists had continued to study only the protein-coding DNA.

THE PURPOSE AND EXTENT
OF NOAH'S FLOOD

S ome Bible interpreters view Noah's Flood (described in Genesis 6–9) as a mass extinction event of global proportions. This interpretation doesn't necessarily arise from the biblical text but more likely from a modern, global perspective of the world. The passage describes an intense 40-day downpour, accompanied by an upsurge from aquifers. The resultant flood wiped out the entire human population of that time and all birds and mammals associated with them—except for the people (Noah and his relatives) and animals aboard the ark.

Both the Genesis account and several other biblical references give significant clues to the Flood's geographical extent.[1] In its lyrical elaboration on the creation days, Psalm 104:6-9 offers an important interpretive guideline:

> You [God] covered it [Earth] with the deep as with a garment;
>> the waters stood above the mountains.
> But at your rebuke the waters fled,
>> at the sound of your thunder they took to flight;
> they flowed over the mountains,
>> they went down into the valleys,
>> to the place you assigned for them.
> You set a boundary they cannot cross;
>> never again will they cover the earth.

The psalmist says water would "never again" inundate the globe—once the continents had formed and the ocean receded (the second creation day[2]). Likewise, Job and Solomon comment that in preparing Earth for humanity, God established a "boundary," "fixed limits," or "doors and bars" for the oceans that "the waters would not overstep."[3]

In reference to Noah's Flood, Peter described how "the world of that time was deluged and destroyed."[4] The Greek phrase used is *tote kosmos*, which literally means "the world at the time when the things under consideration were taking place."[5] In an ancient context, "world" was considered in terms of people rather than in terms of the planet. Peter's words clarify that Noah's world was not the same as that of a later time.

The regional limitation of the Flood may be inferred from the Genesis story itself. Noah's comment in 7:19 that "all the high mountains under the entire heavens were covered" likely implies that from one horizon to the other Noah could see only water. Later, with the receding of the waters well underway, Genesis 8:5 reports that from his vantage point atop the ark Noah could see the tops of distant hills and mountains. The dove he then released could find only "water over all the surface of the earth" (8:9). Perhaps the dove flew too low over the water to see the faraway land that Noah could easily distinguish from his higher point of view.

The most compelling biblical argument for a regional rather than global catastrophe comes from reflection on God's purpose in sending the Flood. An epidemic of evil threatened humanity. As with disease, perhaps close human contact had heightened the contagion of sin. God's subsequent emphasis on the necessity to spread out suggests that prior to the Flood humans had huddled together in one geographical area—a direct violation of God's earlier command to multiply and fill the earth.[6] His judgment against human wickedness—like surgical removal of a deadly cancer—need not extend beyond their geographical boundary.

Even after the Flood, people resisted God's command to move. Genesis 11 explains how God forcibly intervened to scatter humanity throughout all the habitable landmasses of Earth.[7] The Bible indicates that humanity initially occupied a small region of the planet. A 40-day downpour in that area would have been sufficient to rescue humanity from complete self-destruction due to the escalation of evil. However, the resultant flood, because of its relatively brief duration, would not have left any significant geological or archeological evidence.

Nevertheless, anthropology should show a relatively early and rapid occupation of all the habitable continents. According to RTB's creation model, given the relationships among people who survived the Flood (all the men on the ark were blood related to one man, Noah, whereas the women on board may not have shared any close blood relationship), genetic research can be expected to show that the most recent common ancestor for males bottlenecks at a later date than does the most recent common ancestor for females.[8]

PREDICTIONS ARISING FROM FOUR CREATION/EVOLUTION MODELS

A model's accuracy in predicting what researchers will discover as they gather more data and achieve greater understanding provides one of the simplest, cleanest, and least controversial means for testing and evaluating its merit. If all or nearly all of the predictions arising from a particular creation/evolution model are contradicted by future research, then that model can be judged as having failed. However, if the model's predictions prove to be wholly or largely correct, then that model can be judged as viable.

In the tables that follow, the predictions arising from the three best-known creation/evolution models are compared with those that arise from the RTB creation model. The predictions listed are in addition to those already presented in chapter 11. Because variants exist for each of the models compared and contrasted in the tables, some predictions apply to only some of the adherents to that set of models. Also, the brevity of the descriptions and/or the lack of development in some of the models limit the discriminatory power of the listed predictions.

The table entries are by no means exhaustive. Each of the four models makes many more predictions. And these entries are worded with extreme brevity. The intent in this appendix is not to lay out predictive tests in an exhaustive or rigorous manner but rather to illustrate how the predictive testing process can be launched.

TABLE F.1: PREDICTIVE TESTS FOR CREATION/EVOLUTION MODELS (SIMPLE SCIENCES)

Some additional contrasting short-range predictions from the simple sciences for four models

	RTB MODEL	NATURALISM	YOUNG-EARTH	THEISTIC EVOLUTION
3*	As astronomers discover more extrasolar planets and learn more about their parent stars, evidence for the rarity of solar system features that permit advanced life will increase, as will the need for intervention miracles to explain them.	As astronomers discover more extrasolar planets and learn more about parent stars of such planets, they will find increasing evidence that the solar system's characteristics that permit the existence of advanced life are relatively common.	As astronomers discover more extrasolar planets and learn more about parent stars of such planets, they will find increasing evidence both for the rarity of the solar system's characteristics that permit the existence of advanced life and for the extreme youthfulness of all stars and planets.	As astronomers discover more extrasolar planets and learn more about parent stars of such planets, they will find increasing evidence that the solar system's features that permit the existence of advanced life all can be explained by God's actions at the cosmic creation event.
4	Astronomers' measurements of the universe's age will become more accurate, more consistent, and more certainly fixed on about 14 billion years.	Astronomers' measurements of the universe's age will increasingly point to the possibility that the universe in some form may be much older than 14 billion years.	Astronomers' measurements of the universe's age either will become extremely discordant or will quickly drop down to values consistent with about 10,000 years.	Astronomers' measurements of the universe's age will become more accurate, more consistent, and more certainly fixed on about 14 billion years.
5	Evidence for ongoing star and planet formation and for star extinction will increase.	Evidence for ongoing star and planet formation and for star extinction will increase.	Evidence for ongoing star and planet formation and for star extinction will decrease and eventually prove false.	Evidence for ongoing star and planet formation and for star extinction will increase.
6	Evidences for an actual beginning of space and time will grow stronger and more numerous. These evidences will continue to place the beginning of space and time at about 14 billion years ago.	Evidences for an actual beginning of space and time will become weaker and less numerous.	New discoveries will prove that the beginning of space and time took place less than about 10,000 years ago.	Evidences for an actual beginning of space and time will grow stronger and more numerous. These evidences will continue to place the beginning of space and time at about 14 billion years ago.

*Predictive tests 1 and 2 appear in chapter 11 (see pages 185–187).

TABLE F.1: PREDICTIVE TESTS FOR CREATION/EVOLUTION MODELS (SIMPLE SCIENCES, CONTINUED)

	RTB MODEL	NATURALISM	YOUNG-EARTH	THEISTIC EVOLUTION
7	Evidence for fine-tuning in the laws and constants of physics and in the gross cosmic features will become stronger and will not be limited to just those cosmic features that are fixed at the cosmic creation event.	Evidence for fine-tuning in the laws and constants of physics and in the gross features of the universe will become weaker.	Evidence for fine-tuning in the laws and constants of physics and in the gross features of the universe that are dependent on a universe billions of years old will become dramatically weaker.	Evidence for fine-tuning in the laws and constants of physics and in the gross features of the universe will become stronger, but will be limited to just those cosmic characteristics that are fixed at the cosmic creation event.
8	New, more precise cosmic expansion measures will confirm that the universe has been expanding for about 14 billion years and that the expansion rate is extraordinarily (supernaturally) fine-tuned.	New, more precise cosmic expansion measures will show either dramatically less fine-tuning or that the fine-tuning has a natural explanation.	New, more accurate measures of cosmic expansion will show that currently established values are about a million times too low.	New discoveries will confirm fine-tuning in the cosmic expansion rate and that the fine-tuning is limited to tuning cosmic characteristics at the beginning of time.
9	Astronomers and physicists will continue to accumulate evidence that the laws and constants of physics have remained extraordinarily fixed over the past 14 billion years — from the present back to the cosmic creation event.	Astronomers and physicists will discover that radically alternate physics operated when the universe was very young — physics that would eliminate both the need for a transcendent beginning of the universe and the evidence for supernatural design.	Astronomers will discover a major discontinuity in the laws and constants of physics corresponding to a time a few thousand years ago when Adam first sinned and/or when Noah's Flood occurred.	Astronomers and physicists will continue to accumulate evidence that the laws and constants of physics have remained extraordinarily fixed over the past 14 billion years — from the present back to the cosmic creation event.

Table F.1: Predictive Tests for Creation/Evolution Models (Simple Sciences, Continued)

	RTB Model	Naturalism	Young-Earth	Theistic Evolution
10	Both the number of characteristics of the Local Group of galaxies that must be fine-tuned and the degree of fine-tuning in those features to make the existence of advanced life possible will progressively increase as astronomers learn more about the Local Group.	Both the number of characteristics of the Local Group of galaxies that must be fine-tuned and the degree of fine-tuning in those features to make the existence of advanced life possible will progressively decrease as astronomers learn more about the Local Group.	Fine-tuning evidence for the existence of advanced life in the characteristics of the Local Group of galaxies that depend on cosmic or galaxy ages older than 10,000 years will progressively decrease as astronomers learn more about the Local Group.	Only the fine-tuning evidence for the existence of advanced life in the characteristics of the Local Group of galaxies that do not require intervention miracles after the creation event will progressively increase as astronomers learn more about the Local Group.
11	The proximity of galaxies to one another will prove to be proportional to their distance from Earth.	More distant galaxies will prove to be closer together but not necessarily in ways that would demand a single big bang creation event.	Distant galaxies will appear to be no more crowded together than nearby galaxies.	The proximity of galaxies to one another will prove to be proportional to their distance from Earth.
12	New discoveries will continue to increase the evidence that humans exist at the ideal moment in cosmic history to foster advanced civilization and to observe the cosmic creation event and measure cosmic design features.	New discoveries will show that there is nothing especially remarkable about the timing of humanity's arrival in cosmic history.	New discoveries will prove that humans arrived on Earth just 144 hours after the beginning of the universe and the beginning of space and time.	Humanity's existence at the ideal moment in cosmic history will prove to be entirely dependent on the initial conditions of the universe.

TABLE F.1: PREDICTIVE TESTS FOR CREATION/EVOLUTION MODELS (SIMPLE SCIENCES, CONTINUED)

	RTB MODEL	NATURALISM	YOUNG-EARTH	THEISTIC EVOLUTION
13	Both the number of characteristics of the Milky Way Galaxy that must be fine-tuned and the degree of fine-tuning in those features to make the existence of advanced life possible will progressively increase as astronomers learn more about the Milky Way Galaxy.	Both the number of characteristics of the Milky Way Galaxy that must be fine-tuned and the degree of fine-tuning in those features to make the existence of advanced life possible will progressively decrease as astronomers learn more about the Milky Way Galaxy.	Fine-tuning evidence for the existence of advanced life in the characteristics of the Milky Way Galaxy that depend on ages older than 10,000 years will progressively decrease as astronomers learn more about the Milky Way Galaxy.	Only fine-tuning evidence for the existence of advanced life in the Milky Way Galaxy's characteristics that do not require intervention miracles after the cosmic creation event will progressively increase as astronomers learn more about the Milky Way Galaxy.
14	As astronomers learn more about the physical requirements for advanced life, they will find increasing evidence for the anthropic principle inequality.	As astronomers learn more about the physical requirements for advanced life, they will find decreasing evidence for the anthropic principle inequality.	As astronomers learn more about the physical requirements for advanced life, they will find decreasing evidence for the anthropic principle inequality.	No majority position yet developed.
15	Both the number of characteristics of the solar system that must be fine-tuned and the degree of fine-tuning in those features to make the existence of advanced life possible will progressively increase as astronomers learn more about the solar system.	Both the number of characteristics of the solar system that must be fine-tuned and the degree of fine-tuning in those features to make the existence of advanced life possible will progressively decrease as astronomers learn more about the solar system.	Fine-tuning evidence for the existence of advanced life in the characteristics of the solar system that depend on ages older than 10,000 years will progressively decrease as astronomers learn more about the solar system.	Only the fine-tuning evidence for the existence of advanced life in the characteristics of the solar system that do not require intervention miracles after the cosmic creation event will progressively increase as astronomers learn more about the solar system.

Table F.1: Predictive Tests for Creation/Evolution Models (Simple Sciences, Continued)

	RTB Model	Naturalism	Young-Earth	Theistic Evolution
16	Both the number of characteristics of Earth that must be fine-tuned and the degree of fine-tuning in those features to make the existence of advanced life possible will progressively increase as astronomers learn more about the Earth.	Both the number of characteristics of Earth that must be fine-tuned and the degree of fine-tuning in those features to make the existence of advanced life possible will progressively decrease as astronomers learn more about the Earth.	Fine-tuning evidence for the existence of advanced life in the characteristics of Earth that depend on ages older than 10,000 years will progressively decrease as astronomers learn more about the Earth.	Only the fine-tuning evidence for the existence of advanced life in the characteristics of Earth that do not require intervention miracles after the cosmic creation event will progressively increase as astronomers learn more about Earth.
17	Both the number of characteristics of the moon-forming collision event that must be fine-tuned and the degree of fine-tuning in those features to make the existence of advanced life possible will progressively increase as astronomers learn more about the moon-forming event.	Both the number of characteristics of the moon-forming collision event that must be fine-tuned and the degree of fine-tuning in those features to make the existence of advanced life possible will progressively decrease as astronomers learn more about the moon-forming event.	As astronomers learn more about the Earth-Moon system, evidence for the moon-forming collision event will progressively wane.	Only the fine-tuning evidence for the existence of advanced life in the characteristics of the moon-forming collision event that do not require intervention miracles after the cosmic creation event will progressively increase as astronomers learn more about the moon-forming event.
18	Both the number of characteristics of the Late Heavy Bombardment that must be fine-tuned and the degree of fine-tuning in those features to make the existence of advanced life possible will progressively increase as astronomers learn more about the bombardment.	Both the number of characteristics of the Late Heavy Bombardment that must be fine-tuned and the degree of fine-tuning in those features to make the existence of advanced life possible will progressively decrease as astronomers learn more about the bombardment.	As astronomers learn more about the solar system, evidence that the Late Heavy Bombardment occurred either during the flood of Noah and/or at the time of Adam's rebellion against God.	Only the fine-tuning evidence for the existence of advanced life in the characteristics of the Late Heavy Bombardment that do not require intervention miracles after the cosmic creation event will progressively increase as astronomers learn more about the bombardment.

TABLE F.1: PREDICTIVE TESTS FOR CREATION/EVOLUTION MODELS (SIMPLE SCIENCES, CONTINUED)

	RTB MODEL	NATURALISM	YOUNG-EARTH	THEISTIC EVOLUTION
19	Future solar system discoveries will reveal the remains of life on all solar system bodies in proportion to how efficiently meteoritic impacts can transport such remains from Earth and to how well conditions on the respective body allow the preservation of such remains.	Future solar system discoveries will establish the existence of indigenous life (not just transported remains from Earth) on and in many solar system bodies besides Earth.	Future solar system discoveries will reveal that any remains of life found on solar system bodies besides Earth were transported there from Earth less than 10,000 years ago.	Some proponents adopt the RTB position, others the naturalism position.
20	Ongoing research increasingly will demonstrate that the habitable zones for life and intelligent life in particular (cosmic, galactic, and planetary system) are narrow.	Ongoing research increasingly will demonstrate that the habitable zones for life and intelligent life in particular (cosmic, galactic, and planetary system) are broad.	Ongoing research increasingly will demonstrate that the habitable zones for life and intelligent life in particular (cosmic, galactic, and planetary system) are narrow.	Ongoing research increasingly will demonstrate that the habitable zones for life and intelligent life in particular (cosmic, galactic, and planetary system) may be narrow or broad.
21	As astronomers discover more planets, they will find increasing evidence that analogs of the solar system and Earth similar enough to permit the existence of advanced life are either rare or nonexistent.	As astronomers discover more planets, they will find increasing evidence that analogs of the solar system and Earth similar enough to permit the existence of advanced life are common.	As astronomers discover more planets, they will find increasing evidence that analogs of the solar system and Earth similar enough to permit the existence of advanced life are either rare or nonexistent.	As astronomers discover more planets, they will find increasing evidence that analogs of the solar system and Earth similar enough to permit the existence of advanced life may or may not be rare.

TABLE F.1: PREDICTIVE TESTS FOR CREATION/EVOLUTION MODELS (SIMPLE SCIENCES, CONTINUED)

	RTB MODEL	NATURALISM	YOUNG-EARTH	THEISTIC EVOLUTION
22	Evidence will become increasingly more compelling that no laws of physics exist that cause extreme local violations of the thermodynamic laws such that complexity and order spring from simplicity and disorder to such a degree that life spontaneously arises from nonlife.	Evidence will become increasingly more compelling that laws of physics do exist that cause extreme local violations of the thermodynamic laws such that complexity and order spring from simplicity and disorder to such a degree that life spontaneously arises from nonlife.	Evidence will become increasingly more compelling that no laws of physics exist that cause extreme local violations of the thermodynamic laws such that complexity and order spring from simplicity and disorder to such a degree that life spontaneously arises from nonlife.	Some proponents adopt the RTB position, others the naturalism position.

TABLE F.2: PREDICTIVE TESTS FOR CREATION/EVOLUTION MODELS (COMPLEX SCIENCES)

Some additional contrasting short-range predictions from the complex sciences that arise from four models

	RTB Model	Naturalism	Young-Earth	Theistic Evolution
3*	As scientists learn more about the origin of the simplest possible independent life and the laws of physics, they will find increasing evidence against any natural law that spontaneously and instantly self-organizes nonorganic matter into viable organisms.	As scientists learn more about the origin of the simplest possible independent life and the laws of physics, they will find increasing evidence against any natural law that spontaneously and instantly self-organizes nonorganic matter into viable organisms.	As scientists learn more about the origin of the simplest possible independent life and the laws of physics, they will find increasing evidence against any natural law that spontaneously and instantly self-organizes nonorganic matter into viable organisms.	As scientists learn more about the origin of the simplest possible independent life and the laws of physics, they will find increasing evidence for a natural law that spontaneously and instantly self-organizes nonorganic matter into viable organisms.
4	Astrochemical research increasingly will establish the inadequacy of any possible natural source of prebiotics to provide all the chemical building blocks in the necessary concentrations and stabilities for a naturalistic origin of life.	Astrochemists soon will find abundant evidence for vast nearby interstellar reservoirs of concentrated complex stable prebiotics replete with all the prebiotic molecules (sugars, amino acids, nucleotides, lipids, etc.) that life demands.	Astrochemical research increasingly will establish the inadequacy of any possible natural source of prebiotics to provide all the chemical building blocks in the necessary concentrations and stabilities for a naturalistic origin of life.	Astrochemists soon will find abundant evidence for vast nearby interstellar reservoirs of concentrated complex stable prebiotics replete with all the prebiotic molecules (sugars, amino acids, nucleotides, lipids, etc.) that life demands.
5	Astrochemists increasingly will establish that there are no natural reservoirs of concentrated, perfectly homochiral (all oriented with the same handedness) amino acids and sugars.	Astrochemists soon will find abundant evidence for vast nearby interstellar reservoirs of concentrated, perfectly homochiral amino acids and sugars.	Astrochemists increasingly will establish that there are no natural reservoirs of concentrated, perfectly homochiral amino acids and sugars.	Astrochemists soon will find abundant evidence for vast nearby interstellar reservoirs of concentrated, perfectly homochiral amino acids and sugars.

*Predictive tests 1 and 2 appear in chapter 11 (see pages 187–189).

	RTB Model	Naturalism	Young-Earth	Theistic Evolution
6	Geneticists increasingly will establish that the simplest possible independent life-forms are nearly as complex as the simplest independent life on Earth today.	Geneticists soon will find abundant evidence that the simplest possible independent life-forms are orders of magnitude simpler than the simplest independent life on Earth today.	Geneticists increasingly will establish that the simplest possible independent life-forms are nearly as complex as the simplest independent life on Earth today.	No majority position yet established.
7	Scientists increasingly will establish that the origin of life took place within a very narrow window of time billions of years ago.	Scientists soon will find abundant evidence that the time window in which life's origin occurred is orders of magnitude longer than what current data show.	Scientists increasingly will establish that the origin of life took place instantaneously less than 10,000 years ago.	No majority position yet established.
8	When scientists recover from the moon some fossils of Earth's first life (circa 3.8 billion years old), they will find that these fossilized remains are as complex or nearly so and diverse as the earliest fossils of life found on Earth (circa 3.5 billion years old).	When scientists recover from the moon some fossils of Earth's first life (circa 3.8 billion years old), they will find that Earth's first life is reduced to a single species that is far simpler than the simplest known life-forms represented in the terrestrial fossil record.	When scientists recover from the moon the fossils of Earth's first life, they will find strong evidence that such life proves no older than 10,000 years.	No majority position yet established.
9	Research on the components of the Drake Equation increasingly will demonstrate that the probability for the existence of extraterrestrial intelligent life is indistinguishable from zero.	Research on the components of the Drake Equation increasingly will demonstrate that the probability for the existence of extraterrestrial intelligent life is high.	Research on the components of the Drake Equation increasingly will demonstrate that the probability for the existence of extraterrestrial intelligent life is indistinguishable from zero.	Research on the components of the Drake Equation may or may not demonstrate that the probability for the existence of extraterrestrial intelligent life is indistinguishable from zero.
10	Future searches for extraterrestrial intelligent life will continue to produce null results.	Future searches for extraterrestrial intelligent life will eventually produce positive results.	Future searches for extraterrestrial intelligent life will continue to produce null results.	Future searches for extraterrestrial intelligent life may or may not produce positive results.

	RTB MODEL	NATURALISM	YOUNG-EARTH	THEISTIC EVOLUTION
11	Chemists will find increasing evidence for chemical barriers blocking naturalistic pathways to the assembly of nucleobases and sugars into all the life-critical DNA and RNA molecules within realistic time scales.	Chemists soon will find easy naturalistic chemical pathways to the assembly of nucleobases and sugars into all the life-critical DNA and RNA molecules within realistic time scales.	Chemists will find increasing evidence for chemical barriers blocking naturalistic pathways to the assembly of nucleobases and sugars into all the life-critical DNA and RNA molecules within realistic time scales.	Chemists soon will find difficult but not unrealistic naturalistic chemical pathways to the assembly of nucleobases and sugars into all the life-critical DNA and RNA molecules within realistic time scales.
12	Chemists will find increasing evidence for chemical barriers blocking naturalistic pathways to the assembly of amino acids into all the life-critical proteins within realistic time scales.	Chemists soon will find easy naturalistic chemical pathways to the assembly of amino acids into all the life-critical proteins within realistic time scales.	Chemists will find increasing evidence for chemical barriers blocking naturalistic pathways to the assembly of amino acids into all the life-critical proteins within realistic time scales.	Chemists soon will find difficult but not unrealistic naturalistic chemical pathways to the assembly of amino acids into all the life-critical proteins within realistic time scales.
13	Chemists will find increasing evidence for chemical barriers blocking naturalistic pathways to the assembly of lipids and amino acids into life-functional membranes within realistic time scales.	Chemists soon will find easy naturalistic chemical pathways to the assembly of lipids and amino acids into life-functional membranes within realistic time scales.	Chemists will find increasing evidence for chemical barriers blocking naturalistic pathways to the assembly of lipids and amino acids into life-functional membranes within realistic time scales.	Chemists soon will find difficult but not unrealistic naturalistic chemical pathways to the assembly of lipids and amino acids into life-functional membranes within realistic time scales.
14	Biochemists will find increasing evidence for a very complex level of efficient organizational design in the interrelationships and functions of molecules within cells.	Biochemists soon will find that the organizational design of molecules within cells is not nearly as complex or efficient as now understood.	Biochemists will find increasing evidence for a very complex level of efficient organizational design in the interrelationships and functions of molecules within cells.	Biochemists will find that the organization of molecules within cells, though difficult to achieve by naturalistic means, is, nonetheless, realistic without invoking interventionist miracles.

TABLE F.2: PREDICTIVE TESTS FOR CREATION/EVOLUTION MODELS (COMPLEX SCIENCES, CONTINUED)

	RTB Model	Naturalism	Young-Earth	Theistic Evolution
15	Chemists will find increasing evidence for elegant, efficient, and optimal designs of molecules within cells. These designs will continue to prove to match or exceed the quality of man-made machines.	Chemists will find increasing evidence that the designs of molecules within cells are not so elegant, efficient, and optimal as they seem at present and that naturalistic explanations will be found for all the designs.	Chemists will find increasing evidence for elegant, efficient, and optimal designs of molecules within cells. These designs will continue to prove to match or exceed the quality of man-made machines.	Chemists will find that though the design of molecules within cells is difficult to achieve by naturalistic means, realistic pathways for such designs will be found without the need for invoking interventionist miracles.
16	The goal of making a simple life-form in the lab from nonorganic compounds will prove increasingly extreme in its technological demands, expense, and design intricacy.	Soon life will be made in the lab from nonorganic compounds. This success will show that realistic naturalistic pathways for life's origin's must exist.	The goal of making a simple life-form in the lab from nonorganic compounds will prove increasingly extreme in its technological demands, expense, and design intricacy.	The quest to make life in the lab from nonorganic compounds will show that though difficult, realistic means for such assembly in nature could occur without interventionist miracles.
17	Research increasingly will reveal that Earth's first life was diverse, widespread, and abundant. Its complexity will resemble that of the simplest life-forms on Earth today.	Research increasingly will show that Earth's first life was a single life-form existing in just one location. This life-form increasingly will prove far simpler than the simplest life-forms on Earth today.	Research increasingly will show Earth's first life existed no more than 10,000 years ago. Earth's first life will prove to be as complex and diverse as present-day life but without the existence of carnivores and parasites.	No majority position yet developed.
18	Research increasingly will show that the speciation events and abundance levels of life over the past 3.8 billion years have been designed to maximize the quantity, quality, and diversity of biodeposits for the support of human civilization.	Research increasingly will show that there is nothing extraordinary, special, or designed about Earth's biodeposits in the context of providing support for human civilization.	Research increasingly will show that virtually all of Earth's biodeposits were laid down during Noah's Flood.	Research increasingly will demonstrate that no interventionist miracles are necessary to explain any of Earth's biodeposits.

	RTB MODEL	NATURALISM	YOUNG-EARTH	THEISTIC EVOLUTION
19	Research increasingly will show that life has been abundant and widespread on Earth throughout the past 3.8 billion years except for very brief intervals after mass extinction events.	Research increasingly will show that life gradually and slowly became more abundant and widespread after life's origin and after each mass extinction event.	Research increasingly will show that life has been present on Earth for no more than 10,000 years and that only one mass extinction event has occurred, Noah's Flood.	Research increasingly will show that life has been abundant and widespread on Earth throughout the past 3.8 billion years.
20	Research increasingly will show that specific species were removed and introduced at specified epochs throughout the past 3.8 billion years to perfectly compensate for changes in the sun's luminosity.	Research increasingly will show that no specification of either the removal or introduction of species or the timing of such events is necessary to compensate for changes in the sun's luminosity.	Research increasingly will show that no compensation for changes in the sun's luminosity is needed through specified extinction and speciation events because evidence will grow that the sun, Earth, and life has existed for less than 10,000 years.	No majority position yet developed.
21	Research increasingly will show that the quantity and diversity of sulfate-reducing bacteria have been carefully regulated to provide vital-poison metals of the just-right levels and types at the just-right times for advanced life.	Research increasingly will show that no special regulation or design of sulfate-reducing bacteria is necessary to provide advanced life with the levels and types of vital-poison metals it needs.	Research increasingly will show that sulfate-reducing bacteria play no significant role in providing advanced life with the vital-poison metals it needs.	Research increasingly will show that no intervention miracles are needed to explain how sulfate-reducing bacteria provide advanced life with the vital-poison metals it needs.
22	Research increasingly will show that bacteria indeed provided humanity with optimally rich, extensive ore deposits and that the speciation and growth of such bacteria must be specified.	Research increasingly will show that there is nothing extraordinary or special about the ore deposits generated by bacteria.	Research increasingly will show that bacteria played virtually no role in providing humanity with ore deposits.	Research increasingly will show that no intervention miracles are necessary to explain how bacteria provided humanity with ore deposits.

	RTB MODEL	NATURALISM	YOUNG-EARTH	THEISTIC EVOLUTION
23	Research increasingly will show that the symbiosis, organization, extent, and timing of cryptogamic colonies must be highly specified to prepare landmasses for advanced life.	Research increasingly will show that no specification of the symbiosis, organization, extent, and timing of cryptogamic colonies is needed to explain how these colonies prepared the landmasses for advanced life.	Research increasingly will show that cryptogamic colonies played no significant role in preparing the continental landmasses for advanced life.	Research increasingly will show that no intervention miracles are necessary to explain how cryptogamic colonies prepared the continental landmasses for advanced life.
24	Research increasingly will show that specific species were removed and introduced at just-right times to alter the kind and extent of land erosion so as to compensate for changes in the sun's luminosity.	Research increasingly will show that no specification is needed in the removal, introduction, or timing of species to explain how land erosion compensates for changes in the sun's luminosity.	Research increasingly will show that no regulation of land erosion is necessary because other studies will prove that the sun, Earth, and life have not existed for more than about 10,000 years.	Research increasingly will show that no intervention miracles are necessary to explain how various species at various times regulated land erosion so as to compensate for changes in the sun's luminosity.
25	Research increasingly will show that the kind, extent, and duration of photosynthetic life must be highly specified to form, at the just-right time, the optimal amount of free oxygen for advanced life.	Research increasingly will show that no specification is needed either in the kind, extent, or duration of photosynthetic life to form the amount of free oxygen optimal for advanced life.	Research increasingly will show that very little of the free oxygen that supports advanced life came from photosynthetic life.	Research increasingly will reveal that no intervention miracles are needed to explain how the free oxygen supporting advanced life arose from photosynthetic life.
26	Research increasingly will show that natural "disasters" have struck Earth in a manner that is highly fine-tuned to remove the just-right species at the just-right times to compensate for changes in the solar system and prepare Earth for humanity.	New research will establish that there is no significant fine-tuning in the manner in which natural disasters removed certain species at certain times from the terrestrial scene and that such removals play no special role in preparing Earth for humanity.	Research will increasingly establish that only one natural disaster, Noah's Flood, played any significant role in removing species from the terrestrial scene.	New research will show that no miraculous interventions are needed to explain how natural disasters removed the just-right species at the just-right times to compensate for changes in the solar system and prepare Earth for humanity.

Table F.2: Predictive Tests for Creation/Evolution Models (Complex Sciences, Continued)

	RTB Model	Naturalism	Young-Earth	Theistic Evolution
27	Research increasingly will establish that the Cambrian explosion is truly explosive in bringing many phyla suddenly and simultaneously on the terrestrial scene with intact optimal ecological relationships.	Research increasingly will show that the Cambrian explosion (CE) is not at all explosive. Rather, the CE phyla will prove to have gradually and sequentially evolved over a long period of time.	New research will reveal that all of the Cambrian explosion phyla appeared less than 10,000 years ago.	New research will show that no miraculous interventions are necessary to explain the Cambrian explosion, the sudden appearance of most of Earth's animal phyla.
28	Research increasingly will confirm that Earth's biological history and geological processes were optimally designed to provide humanity with the richest possible fossil fuel deposits.	New research will show that no extraordinary fine-tuning is necessary to explain the abundant and rich fossil fuel deposits currently available for humanity.	New research will establish that virtually all of Earth's fossil fuel deposits were laid down during Noah's Flood.	New research will show that no miraculous interventions are necessary to explain the abundant and rich fossil fuel deposits available to humanity.
29	Research will increasingly confirm that explosive mass speciation events have occurred, events in which thousands of new species suddenly appeared on the terrestrial scene without any apparent connection to previously existing species.	New research will show that all mass speciation events have realistic naturalistic explanations. Accordingly, speciation events will prove to be other than mass, sudden, coincident, and widespread.	New research will show that all speciation events occurred within the last 10,000 years. Further, research will prove that the most dramatic mass speciation events occurred right after Adam's rebellion against God and right after Noah's Flood.	New research will demonstrate that no intervention miracles are needed to explain any of the mass speciation events.
30	Research increasingly will confirm that the time interval between some mass extinction events and subsequent mass speciation events is far too brief for any possible naturalistic cause.	Research increasingly will show that the time gaps between all mass extinction events and subsequent mass speciation events are wide enough to permit a reasonable natural cause.	New research will establish that all the mass speciation and extinction events occurred within the last 10,000 years.	New research will show that no intervention miracles are necessary to explain any of the mass speciation or mass extinction events.

TABLE F.2: PREDICTIVE TESTS FOR CREATION/EVOLUTION MODELS (COMPLEX SCIENCES, CONTINUED)

	RTB MODEL	NATURALISM	YOUNG-EARTH	THEISTIC EVOLUTION
31	Research increasingly will show that large-bodied species with small populations and long generation times manifest extinction times far briefer than any epoch during which they could naturally evolve into a distinctly different species.	Research increasingly will show that large-bodied species with small populations and long generation times manifest extinction times longer than the epoch during which they could naturally evolve into a distinctly different species.	New research will prove that the extinction times of all species are irrelevant to creation/evolution issues because such research will establish that Earth is less than 10,000 years old.	New research will establish that no intervention miracles are needed to explain how any of the large-bodied, small-population, long-generation-time species evolved into distinctly different species before they went extinct.
32	Research increasingly will confirm that deleterious mutations, natural disasters, and changes in the terrestrial and solar system environments imply brief extinction times for large-bodied, small-population, long-generation-time species.	Research increasingly will confirm that deleterious mutations, natural disasters, and changes in Earth's and the solar system's environments do not imply brief extinction times for large-bodied, small-population, long-generation-time species.	New research will show that extinction times for large-bodied, small-population, short-generation-time species are irrelevant to creation/evolution issues because such research will prove that Earth is less than 10,000 years old.	No majority position yet established.
33	Research increasingly will show that significant animal speciation, though prolific before the advent of humanity, ceased with the arrival of humanity.	Research will establish that significant animal speciation did not cease or significantly change with the advent of humanity.	Research will prove that no speciation occurred prior to 144 hours before humanity's advent and that mass speciation events took place after Adam's rebellion and after Noah's Flood.	New research will show that no intervention miracles are necessary to explain any of the animal speciation events.

TABLE F.2: PREDICTIVE TESTS FOR CREATION/EVOLUTION MODELS (COMPLEX SCIENCES, CONTINUED)

	RTB MODEL	NATURALISM	YOUNG-EARTH	THEISTIC EVOLUTION
34	Research increasingly will provide examples of repeated optimized designs in species that are distantly or not related from an evolutionary perspective.	Research increasingly will show that examples of repeated designs in species distantly or not related from an evolutionary perspective are either nonexistent or very rare.	Research increasingly will provide examples, though not as many as in the RTB model, of repeated optimized designs in species that are distantly or not related from an evolutionary perspective.	New research will show that no intervention miracles are necessary to explain any of the examples of repeated designs in species that are distantly or not related from an evolutionary perspective.
35	Research increasingly will show that apparent "transitional forms" appear more frequently and rapidly in the fossil record for large-bodied, small-population, long-generation-time species than for small-bodied, large-population, short-generation-time species.	Research increasingly will show that apparent transitional forms appear more frequently and rapidly in the fossil record for small-bodied, large-population, short-generation-time species than for large-bodied, small-population, long-generation-time species.	Research increasingly will show that the only occasions for transitional forms of any type will be brief episodes after Adam's rebellion and after Noah's Flood.	No majority position yet developed.
36	Research increasingly will establish that the probability of humans arising from bacteria even granted optimistic evolutionary assumptions is for all practical purposes zero.	Research increasingly will establish realistic probabilities of humans arising from bacteria through strictly naturalistic means.	Research increasingly will establish that the probability of humans arising from bacteria even granted optimistic evolutionary assumptions is for all practical purposes zero.	Some proponents adopt the RTB position, others the naturalism position.
37	Continuing DNA analysis increasingly will establish that humans could not have naturally descended from previously existing hominids or primates.	Continuing DNA analysis increasingly will establish that humans did naturally descend from previously existing hominids or primates.	Continuing DNA analysis increasingly will establish that Neanderthals, archaic Homo sapiens, and Homo erectus are fully human.	No majority position yet developed.

TABLE F.2: PREDICTIVE TESTS FOR CREATION/EVOLUTION MODELS (COMPLEX SCIENCES, CONTINUED)

	RTB MODEL	NATURALISM	YOUNG-EARTH	THEISTIC EVOLUTION
38	Research increasingly will establish that the entire human race is descended from one man and one woman living in one location not far from the juncture of Africa, Asia, and Europe a few or several tens of thousands of years ago.	Research increasingly will contradict the theory that the entire human race is descended from one man and one woman living in one location not far from the juncture of Africa, Asia, and Europe a few or several tens of thousands of years ago.	Research increasingly will establish that the entire human race is descended from one man and one woman living in one location not far from the juncture of Africa, Asia, and Europe less than 10,000 years ago.	Most theistic evolutionists would concur with the naturalism prediction. Some would agree with the RTB model prediction.
39	Research increasingly will show that humans, and only humans, among all species, past or present, possess a spiritual nature and manifest spiritual capabilities.	Research increasingly will show that many of humanity's spiritual capabilities are shared with other species and all of them can be explained as properties gained through strictly natural descent from earlier species.	Research increasingly will show that humans, and only humans, among all species, past or present, manifest spiritual capabilities and that these capabilities have existed only during the last 10,000 years or less.	No majority position yet established.
40	Research increasingly will show that humans, and only humans, among all species, past or present, possess all the brain structures needed to service spiritual activity.	Research increasingly will show that the human brain is not especially unique in its capacity to service spiritual activity and that any unique features will prove to be minor and explainable by strictly natural evolution.	Research increasingly will show that only humans, Neanderthals, archaic *Homo sapiens*, and *Homo erectus* possess all the brain structures needed to service spiritual activity.	No majority position yet established.

	RTB MODEL	NATURALISM	YOUNG-EARTH	THEISTIC EVOLUTION
41	Research increasingly will establish that no significant change has occurred in either human or Neanderthal DNA over the historical and geographical ranges for both species.	Research increasingly will reveal that significant changes have occurred in both human and Neanderthal DNA over the historical and geographical ranges for both species.	Young-earth creationists hold to a position on this issue somewhere between that of naturalism and the RTB model.	Theistic evolutionists hold to a position on this issue somewhere between that of naturalism and the RTB model.
42	When DNA is recovered and analyzed from *Homo sapiens idaltu* and other archaic *Homo sapiens*, it will prove to be so distinct from human DNA as to rule out these species as natural human ancestors.	When DNA is recovered and analyzed from *Homo sapiens idaltu* and other archaic *Homo sapiens*, it will prove similar enough to human DNA to establish these species as natural human ancestors.	When DNA is recovered and analyzed from *Homo sapiens idaltu* and other archaic *Homo sapiens*, it will prove so similar to human DNA as to establish that these species were fully human.	Most theistic evolutionists would concur with the naturalism prediction. Some would agree with the RTB model prediction.
43	When DNA is recovered and analyzed from *Homo sapiens idaltu* and other archaic *Homo sapiens*, it will show that these species did not experience significant natural evolution.	When DNA is recovered and analyzed from *Homo sapiens idaltu* and other archaic *Homo sapiens*, it will show that these species did experience significant natural evolution.	Young-earth creationists hold to a position on this issue somewhere between that of naturalism and the RTB model.	Theistic evolutionists hold to a position on this issue somewhere between that of naturalism and the RTB model.
44	Research increasingly will show that no significant changes have occurred in the morphology of humans and of any of the hominid species preceding humanity over their historical and geographical ranges.	Research increasingly will show that significant changes have occurred in the morphology of humans and of the hominid species preceding humanity over their historical and geographical ranges.	Young-earth creationists hold to a position on this issue somewhere between that of naturalism and the RTB model.	Theistic evolutionists hold to a position on this issue somewhere between that of naturalism and the RTB model.

Table F.2: Predictive Tests for Creation/Evolution Models (Complex Sciences, Continued)

	RTB Model	Naturalism	Young-Earth	Theistic Evolution
45	Research increasingly will show that chimpanzee and human DNA are sufficiently dissimilar to rule out a common recent ancestor through natural descent.	Research increasingly will show that chimpanzee and human DNA are sufficiently similar to establish a common recent ancestor through natural descent.	Research increasingly will show that chimpanzee and human DNA are sufficiently dissimilar to rule out a common recent ancestor through natural descent.	Theistic evolutionists hold to a position on this issue somewhere between that of naturalism and the RTB model.
46	Research increasingly will show that several dramatic cultural revolutions coincide with the advent of humanity.	Research increasingly will discredit the theory that several dramatic cultural revolutions coincide with the advent of humanity.	Research increasingly will show that several dramatic cultural revolutions coincide with the advent of humanity and that this advent occurred less than 10,000 years ago.	Theistic evolutionists hold to a position on this issue somewhere between that of naturalism and the RTB model.
47	Research increasingly will show that humans were endowed at their advent with capabilities for high-tech global civilization that were costly to their initial survival.	Research increasingly will show that all of humanity's initial endowments were vital for their early survival.	Research increasingly will show that humans were endowed at their advent with capabilities for high-tech global civilization that were costly to their initial survival.	Theistic evolutionists hold to a position on this issue somewhere between that of naturalism and the RTB model.
48	Research increasingly will establish a very narrow window of time within cosmic history during which an intelligent physical species could exist.	Research increasingly will show a large fraction of time within cosmic history during which an intelligent physical species could exist.	Research increasingly will show that the window of time during which an intelligent physical species could exist is virtually identical to the entire (young) age of the universe.	No majority position yet established.

TABLE F.2: PREDICTIVE TESTS FOR CREATION/EVOLUTION MODELS (COMPLEX SCIENCES, CONTINUED)

	RTB Model	Naturalism	Young-Earth	Theistic Evolution
49	Research increasingly will establish an extremely narrow window of time within cosmic history during which an intelligent physical species manifesting high technology could exist.	Research increasingly will show a rather broad window of time within cosmic history during which an intelligent physical species manifesting high technology could exist.	Research increasingly will show that the time window during which an intelligent physical species manifesting high-tech civilization could exist is virtually identical to the entire age of the universe.	No majority position yet established.
50	Research increasingly will show that apparently bad or useless designs in nature's complex creatures, once fully understood, are really elegant, efficient, and optimal designs.	Research increasingly will show that bad designs in nature's complex creatures are ubiquitous and that the later a creature appears and the more complex it is the higher its probability to possess bad designs.	Research increasingly will show that apparently bad designs in nature's complex creatures, once fully understood, are either good designs or the results of Adam's rebellion against God.	Theistic evolutionists hold to a position on this issue significantly closer to the RTB model than to the naturalism model.
51	Research increasingly will confirm that natural "disasters" are optimally designed to protect and provide for life's needs and for humanity's in particular.	Research increasingly will demonstrate that natural disasters need not be fine-tuned or designed to protect and provide for life's requirements and for humanity's in particular.	Research increasingly will reveal that natural disasters did not happen until after Adam's rebellion; that is, natural disasters are part of the curse God pronounced upon the earth and life.	Theistic evolutionists hold to a position on this issue significantly closer to the RTB model than to the naturalism model.
52	Research increasingly will confirm that very little of the genomes contains useless or junk DNA, that is, DNA that serves no purpose for the organism.	Research increasingly will show that genomes are filled with DNA portions that serve no purpose and that the more advanced the species and later it appears the more junk its genome contains.	Research increasingly will confirm that very little of the genomes contains useless or junk DNA, that is, DNA that serves no purpose for the organism.	Theistic evolutionists hold to a position somewhere between the naturalism model and the RTB model.

TABLE F.3: PREDICTIVE TESTS FOR CREATION/EVOLUTION MODELS (THEOLOGY/PHILOSOPHY)
Some contrasting short-range predictions from theology and philosophy for four models

	RTB Model	Naturalism	Young-Earth	Theistic Evolution
1	As theologians more thoroughly integrate all 22 of the Bible's major creation accounts and other creation verses, they will discover increasing biblical consistency in the old-earth creationist view and increasing discordance in the young-earth view and in the time scales for the emergence of creation events demanded by theistic evolution.	The Bible's 22 major creation accounts and remaining creation verses will, with advancing theological research, prove to be contradictory, at least with respect to the record of nature.	As theologians more thoroughly integrate all 22 of the Bible's major creation accounts and other creation verses, they will discover increasing biblical consistency for the young-earth view and increasing discordance for the old-earth positions.	As theologians more thoroughly integrate all 22 of the Bible's major creation accounts and other creation verses, they will discover increasing biblical consistency for the old-earth view and for gradual transitions between creation events rather than their abrupt occurrences.
2	A careful study of creation verbs and nouns in the Bible will show that many of the creation events beyond the cosmic beginning are abrupt interventionist miracles.	A study of the Bible's creation verbs and nouns either will show none of the "creation events" to be miraculous or that the Bible's creation content and order contradicts the established record of nature.	A careful study of creation verbs and nouns in the Bible will show that all of the creation events beyond the cosmic beginning are abrupt interventionist miracles.	A careful study of creation verbs and nouns in the Bible will show that none of the creation events beyond the cosmic beginning is an abrupt interventionist miracle.
3	Biblical inerrancy will become more defensible for an old-earth position, with many abrupt interventionist creation miracles beyond the cosmic creation event, than it will for any other creation/evolution view.	Biblical inerrancy will prove increasingly impossible to defend for the biblical passages pertaining to creation.	Biblical inerrancy will become more internally and externally defensible with the young-earth view than with any of the old-earth positions.	Biblical inerrancy will become more defensible for an old-earth position without abrupt interventionist creation miracles besides the cosmic creation event than it will be for any other creation/evolution view.

	RTB MODEL	NATURALISM	YOUNG-EARTH	THEISTIC EVOLUTION
4	As theologians more thoroughly integrate all the creation accounts and verses in the Bible, they will find increasing evidence that Adam and Eve's creation date falls between 10,000 and 100,000 years ago.	As honest theologians more thoroughly integrate all the creation accounts and verses in the Bible, they will find increasing evidence that its content on human origins lacks credibility.	As theologians more thoroughly integrate all the Bible's creation accounts and verses, they will find increasing evidence that Adam and Eve's creation date falls between 6,000 and 10,000 years ago.	As theologians more thoroughly integrate all the creation accounts and verses in the Bible, they will find increasing evidence that Adam and Eve's creation date exceeds a half million years ago.
5	As theologians more thoroughly integrate all the creation accounts and verses in the Bible, they will find increasing biblical evidence for a one-Adam, one-Eve model.	As theologians more thoroughly integrate all the creation accounts and verses in the Bible, they may find increasing biblical confusion about the identities and timing of Adam and Eve.	As theologians more thoroughly integrate all the creation accounts and verses in the Bible, they will find increasing biblical evidence for a one-Adam, one-Eve model.	As theologians more thoroughly integrate the Bible's creation content, they will find increasing biblical evidence for a two- or three-Adam and a two- or three-Eve model.
6	As theologians more thoroughly integrate the Bible's creation content, they will find increasing biblical evidence that a local flood wiped out all humans and their birds and mammals except for those on Noah's ark.	As theologians more thoroughly integrate all the creation accounts and verses in the Bible, they will find increasing chaos and discordance about what the accounts and verses say about Noah's Flood relative to the scientific evidence.	As theologians more thoroughly integrate all the creation accounts and verses in the Bible, they will find increasing biblical evidence for a global flood that wiped out all land-dwelling life except for the life on Noah's ark.	As theologians more thoroughly integrate all the creation accounts and verses in the Bible, they will find increasing evidence for figurative and/or metaphorical interpretations of Noah's Flood.

Table F.3: Predictive Tests for Creation/Evolution Models (Theology/Philosophy, Continued)

	RTB Model	Naturalism	Young-Earth	Theistic Evolution
7	The apparently universal human drive to pursue ultimate purpose and destiny will indeed prove to be real and unique to the human species without any evidence of having evolved over the human era.	The apparently universal human drive to pursue ultimate purpose and destiny will prove either a myth or simply a means to enhance physical survival.	The apparently universal human drive to pursue ultimate purpose and destiny will indeed prove to be real, ubiquitous, and unique to the human species without any evidence of having evolved over the human era.	The apparently universal human drive to pursue ultimate purpose and destiny will prove real and probably will prove to have evolved through the hominids to modern humans.
8	As philosophers continue to research humanity's nature, they will discover increasing logical evidence that humans possess many features that could not possibly be derived or inherited from other animals.	As philosophers continue to research the nature of humanity, they will discover increasing logical evidence that humans possess only features that could possibly be naturally derived or inherited from other animals.	As philosophers continue to research the nature of humanity, they will discover increasing philosophical evidence that humans possess many features that could not possibly be derived or inherited from other animals.	As philosophers continue to research humanity's nature, they will find increasing philosophical evidence that humans possess few, if any, features that could not be derived or inherited from other animals.
9	As philosophers continue to research the nature of birds and mammals, they will find increasing philosophical evidence that they possess many features that could not possibly be derived or inherited from lower animals.	As philosophers continue to research the nature of birds and mammals, they will discover increasing philosophical evidence that they possess no features that could not possibly be derived or inherited from lower animals.	As philosophers continue to research the nature of birds and mammals, they will find increasing philosophical evidence that they possess many features that could not possibly be derived or inherited from lower animals.	As philosophers research the nature of birds and mammals, they will find increasing philosophical evidence they possess no features that could not be derived or inherited from lower animals.
10	As theologians more thoroughly integrate all the Bible's creation accounts and verses, they will find increasing biblical indications of God's plan to place redeemed humans in a new creation that is radically distinct from Eden.	As theologians more thoroughly integrate all the creation accounts and verses in the Bible, they will discover that the Christian doctrines of both paradise and a new creation are completely untenable.	As theologians more thoroughly integrate all the Bible's creation accounts and verses, they will find increasing biblical evidence that God will place redeemed humans in a restored Eden-like paradise on the present Earth.	No majority opinion position yet developed.

TABLE F.3: PREDICTIVE TESTS FOR CREATION/EVOLUTION MODELS (THEOLOGY/PHILOSOPHY, CONTINUED)

	RTB MODEL	NATURALISM	YOUNG-EARTH	THEISTIC EVOLUTION
11	Research will cause philosophical arguments for the existence of the God of the Bible to become progressively stronger.	Research will progressively overthrow all the philosophical arguments for the existence of the God of the Bible.	Research will cause philosophical arguments for the existence of the God of the Bible to become progressively stronger.	Research will cause most, but not all, of the philosophical arguments for the existence of the God of the Bible to become progressively stronger.
12	As theologians more thoroughly integrate all the Bible's creation accounts and verses, they will find increasing biblical evidence for a literal and consistent interpretation of the Bible's creation content.	As theologians more thoroughly integrate all the Bible's creation accounts and verses, they will find increasing difficulties and contradictions for any literal interpretation of the Bible's creation content relative to the scientific record of nature.	As theologians more thoroughly integrate all the Bible's creation accounts and verses, they will find increasing biblical evidence for a literal interpretation of the Bible's creation content assuming that the poetic creation texts deal only with the present creation.	As theologians more thoroughly integrate all the Bible's creation accounts and verses, they will find increasing biblical evidence for metaphorical and figurative interpretations of the Bible's creation content.
13	Ongoing philosophical and theological research will find more redemptive analogies in the Bible and in human culture, history, and experience.	Ongoing philosophical and theological research will prove that more and more redemptive analogies in the Bible and in human culture, history, and experience are inaccurate.	Ongoing philosophical and theological research will produce more redemptive analogies in the Bible and in human culture, history, and experience.	Ongoing philosophical and theological research may or may not produce more redemptive analogies in the Bible and in human nature and culture.
14	Ongoing philosophical and theological research will show that redemptive analogies in the Bible and in human culture, history, and experience more closely resemble the biblical message of redemption than previously thought.	Ongoing philosophical and theological research will show that "redemptive analogies" in the Bible and in human culture, history, and experience prove a poorer fit with the biblical message of redemption than previously thought.	Ongoing philosophical and theological research will show that redemptive analogies in the Bible and in human culture, history, and experience more closely resemble the biblical message of redemption than previously thought.	Ongoing philosophical and theological research will show redemptive analogies in the Bible and in human culture and experience may or may not fit more closely to the biblical message of redemption than previously thought.

TABLE F.3: PREDICTIVE TESTS FOR CREATION/EVOLUTION MODELS (THEOLOGY/PHILOSOPHY, CONTINUED)

	RTB Model	Naturalism	Young-Earth	Theistic Evolution
15	Future research on the Bible's creation texts increasingly will show that the Bible correctly predicted the continuous expansion and many other characteristics of the universe.	Future research on the Bible's creation texts increasingly will show that the Bible failed to predict the continuous expansion or any other characteristics of the universe.	Future research on the Bible's creation texts increasingly will show that the Bible correctly predicted some of the expansion characteristics of the universe.	No majority opinion position yet developed.

NOTES

Introduction: Sifting Fact from Fiction

1. *The War of the Worlds: Mars' Invasion of Earth, Inciting Panic and Inspiring Terror from H. G. Wells to Orson Welles and Beyond* (Naperville, IL: Sourcebooks MediaFusion, 2005), 9.
2. *War of the Worlds*, 12.
3. "New Image of Infant Universe Reveals Era of First Stars, Age of Cosmos, and More," http://www.nasa.gov/centers/goddard/news/topstory/2003/0206mapresults .html, accessed February 11, 2003.
4. C. J. Allègre, G. Manhès, and C. Göpel, "The Age of the Earth," *Geochemica et Cosmochemica Acta* 59 (1995), 1445–1456.
5. Aristotle, *Metaphysics* 3.2.996b28-31.
6. Hugh Ross, *The Creator and the Cosmos: How the Greatest Scientific Discoveries of the Century Reveal God*, 3rd ed. (Colorado Springs, CO: NavPress, 2001); Fazale Rana and Hugh Ross, *Origins of Life: Biblical and Evolutionary Models Face Off* (Colorado Springs, CO: NavPress, 2004); Fazale Rana with Hugh Ross, *Who Was Adam? A Creation Model Approach to the Origin of Man* (Colorado Springs, CO: NavPress, 2005); Hugh Ross, *A Matter of Days: Resolving a Creation Controversy* (Colorado Springs, CO: NavPress, 2004); Hugh Ross, Kenneth Samples, and Mark Clark, *Lights in the Sky and Little Green Men: A Rational Christian Look at UFOs and Extraterrestrials* (Colorado Springs, CO: NavPress, 2002).

Chapter 1: The Creation/Evolution Battlefield

1. *The War of the Worlds: Mars' Invasion of Earth, Inciting Panic and Inspiring Terror from H. G. Wells to Orson Welles and Beyond* (Naperville, IL: Sourcebooks MediaFusion, 2005), 23.
2. *War of the Worlds*, 26.
3. *War of the Worlds*, 22.
4. Edward J. Larson and Larry Witham, "Scientists Are Still Keeping the Faith," *Nature* 386 (1997): 435–436.
5. See the Reasons To Believe Web site for a list of those who hold to an old-earth view. "Notable Christians Open to an Old Earth Interpretation," http://www.reasons.org/

resources/apologetics/notable_leaders/index.shtml, accessed October 11, 2005.

6. Delos B. McKown, *The Myth Maker's Magic: Behind the Illusion of "Creation Science"* (Buffalo, NY: Prometheus Books, 1993), 31; Delos B. McKown, "Unwanted Knowledge," *Free Inquiry* 8, no. 3 (Summer 1988): 22.

7. One example occurred during a lecture I gave on the latest scientific evidences for the existence of the Bible's God to students and their teachers at a Christian high school. Following my message, one teacher insisted I was a theistic evolutionist and, therefore, I had departed from the Christian faith. My protestations did nothing to dissuade him or most of his peers. Later, that school sent out a letter to the parents of its students apologizing for inviting me to speak and explicitly denying my "false teachings."

8. Ian Plimer, *Telling Lies for God: Reason vs. Creationism* (Milsons Point, New South Wales, Australia: Random House Australia, 1994), 10.

9. Several years ago, a writer for a major national television network interviewed me for a documentary on the creation/evolution controversies. After 45 minutes of telephone conversation, the writer said that although my positions on the issues were very interesting, in all likelihood his network would decline the interview. When pressed for more specific reasons, the writer said that unlike other creationists they were interviewing, my positions were too logical and reasonable. He explained that his bosses wanted fireworks, not resolution.

10. An example of this type of branding, one of many, occurred during an evening lecture I gave to students and faculty at the University of Alberta, Calgary campus. After speaking for 30 minutes on the latest scientific evidence, primarily from astronomy and physics—for the God of the Bible—one of the faculty (an atheist quantum physicist) gave a 25-minute response. He told the audience that I was an evangelical Christian who interpreted the Bible literally. Thus, he insisted I must have lied when I said I believe the universe is about 15 billion years old, that Earth is about 4.5 billion years old, and that life has existed on Earth for over 3.5 billion years. Rather than respond to any of my scientific evidence, the quantum physicist spent the remainder of his speaking time refuting young-earth science and the young-earth interpretation of the Bible.

11. J. P. Moreland, "What Is Truth and Why Does It Matter?" http://boundless.org/features/a0000911.html, accessed October 28, 2004.

Chapter 2: Battle Plans: Who's Saying What?

1. Niles Eldredge, *The Triumph of Evolution and the Failure of Creationism* (New York: W. H. Freeman, 2000).

2. J. H. Leuba, *The Belief in God and Immortality: A Psychological, Anthropological, and Statistical Study* (Boston: Sherman, French, and Co., 1916).

3. Humphrey Taylor, "The Religious and Other Beliefs of Americans 2003," The Harris Poll #11, February 26, 2003, http://www.harrisinteractive.com/harris_poll/index.asp?PID=359, accessed August 15, 2005.

4. Frank Newport, "Third of Americans Say Evidence Has Supported Darwin's Evolution Theory," http://www.gallup.com/poll/content/login.aspx?ci=14107, accessed November 2004.

5. "Nearly Two-Thirds of U.S. Adults Believe Human Beings Were Created by God," The Harris Poll #52, July 6, 2005, http://harrisinteractive.com/harris_poll/index.asp?PID=581, accessed December 19, 2005.

6. Peter Schrag, "It's Time for Equal Time for Darwinian Evolution," *Sacramento Bee*, August 10, 2005, http://www.sacbee.com/content/politics/columns/schrag/story/13387051p-14228586c.html, accessed August 15, 2005.

7. Edward J. Larson and Larry Witham, "Scientists Are Still Keeping the Faith," *Nature* 386 (1997): 435–436. In a follow-up study Larson and Witham surveyed members of the U.S. National Academy of Sciences [Edward J. Larson and Larry Witham, "Leading Scientists Still Reject God," *Nature* 394 (1998): 313]. There they found a much lower level of belief in God and immortality and a level of belief significantly lower than Leuba's more selected survey of "leading scientists." However, only members of the USNAS can elect new members, and most of America's leading scientists are not members. Therefore, one would be mistaken to conclude, based on Larson and Witham's second study, that today the best scientists manifest much less belief in God and immortality than they did in 1916.

8. Edward O. Wilson, ed., *From So Simple a Beginning: The Four Great Books of Charles Darwin* (New York: Norton, 2006), 1479.

9. The number of professional scientists and science professors who believe (at least publicly) in the foundational young-earth creationist doctrines of a recent global Earth-altering flood and ages for the universe and Earth in the thousands rather than billions of years based on scientific evidence alone (that is, independent of any particular interpretation of the Bible or any other holy book) has been noted by several Christian organizations to measure zero. Even prominent young-earth creationist leaders admit this assessment is correct. For example, talk radio host John Stewart asked John Morris, president of the Institute for Creation Research, in my presence and broadcast on the air (*Bible on the Line*, KKLA, December 6, 1987), if he or any of his associates had ever met or heard of a scientist who became persuaded the earth or the universe is only thousands of years old based on scientific evidence, without any reference to a particular interpretation of the Bible. Morris answered no. Stewart has since asked the same question of several other prominent young-earth creationist proponents, and the answer has been a consistent no. Likewise, I have continued asking this question of young-earth creationist leaders with the same result.

10. While a few professors teaching at conservative evangelical seminaries do believe in a recent global flood and ages for the earth and universe less than a few tens of thousands of years, I and others have observed that their numbers represent a rapidly shrinking minority. My conclusion is based on my experience in addressing professors at leading conservative evangelical seminaries over the past 25 years where in each instance I asked the professors in attendance to estimate the percentage of their colleagues that would uphold one or both of these two young-earth creationist doctrines.

11. *Edwards v. Aguillard*, 482 U.S. 578 (1978), statement 1(a) under "Held," http://caselaw.lp.findlaw.com/cgi-bin/getcase.pl?court=us&vol=482&invol=578, accessed August 26, 2003.

12. This point is conceded even by the nontheistic skeptic and anticreationist Michael Shermer in his book *Why People Believe Weird Things: Pseudoscience, Superstitions, and Other Confusions of Our Times* (New York: W. H. Freeman, 1997), 162.

13. No major Christian denomination, either pre-Christopher Columbus or post-Christopher Columbus, has ever held to a flat-earth doctrine. However, several churches and a few small sects did teach that Earth was flat. Perhaps the most

famous flat-earth proponent was the radio preacher Wilbur Glenn Voliva of Zion, Illinois. For a list of flat-earth advocates see "Science Reference Guides: The Flat Earth and Its Advocates: A List of References," http://www.loc.gov/rr/scitech/ SciRefGuides/flatearth.html, accessed April 4, 2004. The Flat Earth Society, for its part, has always claimed that the Bible is a flat-earth book. I heard this claim myself years ago when I participated in a radio debate on the shape of Earth on KKLA.

14. Prominent Protestant apologists for geocentrism include Gerardus Bouw, Malcolm Bowden, James Hanson, and Tom Willis. Leading Catholic apologists for geocentrism are R. G. Elmendorf, Paul Ellwanger, Paula Haigh, and Robert Sungenis. The two largest nonprofit organizations dedicated to promoting geocentrism are Catholic Apologetics International and The Biblical Astronomer. The following Web sites are devoted to promoting geocentrism: The Biblical Astronomer at http://www .geocentricity.com; Geocentrism Bulletin Board at http://www.fixedearth.com; and Catholic Apologetics International at http://www.catholicintl.com., accessed April 4, 2004.

15. Phillip E. Johnson, *Darwin on Trial*, 2nd ed. (Downers Grove, IL: InterVarsity, 1993); Phillip E. Johnson, *Reason in the Balance: The Case Against Naturalism in Science, Law, and Education* (Downers Grove, IL: InterVarsity, 1995); Phillip E. Johnson, *Defeating Darwinism by Opening Minds* (Downers Grove, IL: InterVarsity, 1997); Phillip E. Johnson, *The Wedge of Truth: Splitting the Foundations of Naturalism* (Downers Grove, IL: InterVarsity, 2000).

16. Phillip E. Johnson, interview by Tal Brooke, "The Intelligent Design Movement: Asking the Right Questions," *Spiritual Counterfeits Project Journal* 27, nos. 2–3 (2003): 11.

17. Robert T. Pennock, "Reply: Johnson's Reason in the Balance," in *Intelligent Design Creationism and Its Critics: Philosophical, Theological, and Scientific Perspectives*, ed. Robert T. Pennock (Cambridge, MA: MIT Press, 2001), 107.

18. Jerry A. Coyne, "Creationism by Stealth," *Nature* 410 (2001): 745–746.

19. Mark Perakh, "A Presentation Without Arguments: How William Dembski Defeats Skepticism, or Does He?" http://members.cox.net/perakm/Dem_burbank21jun02 .htm, accessed April 2, 2006.

20. Lawrence M. Krauss, "'Creationism' Discussion Belongs in Religion Class," *Reports of the National Center for Science Education* 22, nos. 1–2 (2002): 11.

21. Jennifer Palonus, "Ohio's Saga Approaches an Intermission," http://crevo .bestmessageboard.com/vThreadID=227, accessed November 15, 2001.

22. The financial support from wealthy young-earth creationist advocates is well-known. An executive at Illustra Media, which produces most of the DVD documentaries for the Intelligent Design movement (IDM), informed me in 2004 that purchases by young-earth creationist organizations exceeded those of all other organizations by a ratio of 19 to 1.

23. Philip Kitcher, "Born-Again Creationism," in *Intelligent Design Creationism and Its Critics*, 271.

24. Adrian L. Melott, "Intelligent Design Is Creationism in a Cheap Tuxedo," *Physics Today* 55 (June 2002): 48–50, http://www.physicstoday.org/vol-55/iss-6/p48a.html, accessed December 28, 2004.

25. Carl Wieland, "AiG's Views on the Intelligent Design Movement," http://www .answersingenesis.org/docs2002/0830_IDM.asp, accessed January 9, 2004.

26. Wieland, http://www.answersingenesis.org/docs2002/0830_IDM.asp.

27. Wieland, http://www.answersingenesis.org/docs2002/0830_IDM.asp.

28. This debate was hosted by William F. Buckley Jr. on the PBS program *Firing Line* on December 29, 1997.
29. I asked this question in a private conversation with a prominent IDM leader in 2003. He acknowledged the need for a model but said that he and his colleagues would need at least eight more years to develop one.
30. *American Heritage Dictionary of the English Language*, 4th ed., s.v. "Model."
31. Patricia Princehouse, "Ohio Overthrows Scopes Legacy," *Reports of the National Center for Science Education* 22, no. 5 (2002): 4.
32. *Tammy Kitzmiller et al. v. Dover Area School District et al.*, Case 4:04-cv-02688-JEJ, December 20, 2005, 43, http://www.pamd.uscourts.gov/kitzmiller/kitzmiller_342 .pdf, accessed February 22, 2006.
33. This new definition of theistic evolution differs from one in common use a century or two ago. Then the term referred to those who believed that life has changed dramatically over the earth's history through a combination of natural processes and divine miraculous acts. See B. B. Warfield, "Creation, Evolution, and Mediate Creation," in *Evolution, Science, and Scripture: Selected Writings*, ed. Mark A. Noll and David N. Livingstone (Grand Rapids, MI: Baker, 2000), 197–210.
34. Stuart Kauffman, *Investigations* (New York: Oxford University Press, 2000), 49–209.
35. I give several examples with documentation in *The Creator and the Cosmos: How the Greatest Scientific Discoveries of the Century Reveal God*, 3rd ed. (Colorado Springs, CO: NavPress, 2001), 119–136, 162–174.
36. Stephen Jay Gould, "Nonoverlapping Magisteria," *Natural History* 106 (March 1977): 16–22.
37. John A. Moore, *From Genesis to Genetics: The Case of Evolution and Creationism* (Berkeley, CA: University of California Press, 2002), 198.
38. Moore, 198.
39. Moore, 198.
40. National Academy of Sciences of the United States of America, "Frequently Asked Questions About Evolution and the Nature of Science," *California Journal of Science Education* 2, no. 1 (Fall 2001): 141, 142.
41. National Academy of Sciences, 141.
42. National Academy of Sciences, 142.
43. See Kenneth Richard Samples, *Without a Doubt: Answering the 20 Toughest Faith Questions* (Grand Rapids, MI: Baker, 2004), 24–26, 190–194.
44. Norman L. Geisler and J. Kerby Anderson, *Origin Science: A Proposal for the Creation-Evolution Controversy* (Grand Rapids, MI: Baker, 1987).
45. Jeremiah 23:24.
46. Psalms 19:1-4; 50:6; 97:6.
47. Jeremiah 33:25.
48. Psalms 119; 19:1-4.
49. Numbers 23:19; Psalms 12:6; 19:7-8; 119:160; Proverbs 30:5; John 10:35; Hebrews 6:18.
50. 1 Thessalonians 5:21.

Chapter 3: Applying the Scientific Method
1. Richard A. Kerr, "Newfound 'Tenth Planet' Puts Pluto Behind the Eight Ball," *Science* 309 (2005): 859.
2. A more detailed account is given in Hugh Ross, *The Fingerprint of God: Recent*

Scientific Discoveries Reveal the Unmistakable Identity of the Creator, 2nd ed. (New Kensington, PA: Whitaker House, 2000), 9–118.

3. See Ross, *The Fingerprint of God*, 39–59, for a review of the history of the development of relativity replete with the relevant equations, anomalies that were resolved, predictions made, and citations to the discovery papers.

4. Lincoln Barnett, *The Universe and Dr. Einstein* (New York: William Sloane Associates, 1948), 106.

5. Albert Einstein, "Erklärung der Perihelbewegung des Merkur aus der allgemeinen Relativitätstheorie," *Sitzungsberichte der Königlich Preussischen Akademie der Wissenschaften* (November 18, 1915): 931–839; Albert Einstein, "Die Grundlage der allgemeinen Relativitätstheorie," *Annalen der Physik* 49 (1916): 769–822. The English translation is in *The Principle of Relativity* by H. A. Lorentz, A. Einstein, H. Minkowski, and H. Weyl with notes by A. Sommerfeld and translated by W. Perrett and G. B. Jeffrey (London: Methuen and Co., 1923), 109–164.

6. Arthur S. Eddington, "The End of the World: From the Standpoint of Mathematical Physics," *Nature* 127 (1931): 450.

7. Arthur S. Eddington, "On the Instability of Einstein's Spherical World," *Monthly Notices of the Royal Astronomical Society* 90 (1930): 668–678.

8. Eddington, "Instability of Einstein's Spherical World," 668–678.

9. Masataka Fukugita and P. J. E. Peebles, "The Cosmic Energy Inventory," *Astrophysical Journal* 616 (2004): 643–668.

10. Fukugita and Peebles, 643–668.

11. Eddington, "End of the World," 450.

12. Eddington, "Instability of Einstein's Spherical World," 672.

13. James H. Jeans, *Astronomy and Cosmogony*, 2nd ed. (Cambridge, UK: Cambridge University Press, 1929), 421.

14. Fred Hoyle, "A New Model for the Expanding Universe," *Monthly Notices of the Royal Astronomical Society* 108 (1948): 372.

15. Hermann Bondi, *Cosmology*, 2nd ed. (Cambridge, UK: Cambridge University Press, 1960), 140.

16. John Gribbin, "Oscillating Universe Bounces Back," *Nature* 259 (1976): 15–16.

17. R. H. Dicke et al., "Cosmic Black-Body Radiation," *Astrophysical Journal Letters* 142 (1965): 415.

18. P. J. E. Peebles, "The Mean Mass Density of the Universe," *Nature* 321 (1986): 27.

19. Arno Penzias in *Cosmos, Bios, and Theos*, ed. Henry Morgenau and Roy Abraham Varghese (La Salle, IL: Open Court, 1992), 83.

20. Stephen W. Hawking, *A Brief History of Time: From the Big Bang to Black Holes* (New York: Bantam, 1988), 127.

21. Fang Li Zhi and Lu Shu Xian, *Creation of the Universe*, trans. T. Kiang (Teaneck, NJ: World Scientific, 1989), 173.

22. *Ecumenical Creeds and Reformed Confessions* (Grand Rapids, MI: CRC Publications, 1988), 15, 22–23, 79, 87–88, 91; *Westminster Assembly of Divines, Westminster Confession of Faith* (Norcross, GA: Great Commission Publications, 1992), 15–16; Hugh Ross, *A Matter of Days: Resolving a Creation Controversy* (Colorado Springs, CO: NavPress, 2004), 51–57.

23. Hugh Ross, "Scriptures Related to Creation," http://www.reasons.org/resources/apologetics/p0014.shtml, accessed August 8, 2005.

24. Hugh Ross, *The Creator and the Cosmos: How the Greatest Scientific Discoveries of the*

Century Reveal God, 3rd ed. (Colorado Springs, CO: NavPress, 2001); Fazale Rana and Hugh Ross, *Origins of Life: Biblical and Evolutionary Models Face Off* (Colorado Springs, CO: NavPress, 2004); Fazale Rana with Hugh Ross, *Who Was Adam? A Creation Model Approach to the Origin of Man* (Colorado Springs, CO: NavPress, 2005); Hugh Ross, Kenneth Samples, and Mark Clark, *Lights in the Sky and Little Green Men: A Rational Christian Look at UFOs and Extraterrestrials* (Colorado Springs, CO: NavPress, 2002); Ross, *A Matter of Days*.

25. Ross, *Fingerprint of God*, 39–118; Ross, *Creator and the Cosmos*, 23–136; Hugh Ross, *Beyond the Cosmos: What Recent Discoveries in Astrophysics Reveal About the Glory and Love of God*, 2nd ed. (Colorado Springs, CO: NavPress, 1999), 27–52; Hugh Ross, "A Beginner's—and Expert's—Guide to the Big Bang," *Facts for Faith*, no. 3 (Q3 2000), 14–32.

26. Ross, *Fingerprint of God*, 107–118; Ross, *Creator and the Cosmos*, 99–136; Ross, *Beyond the Cosmos*, 27–46; Hugh Ross, "Cosmic Scans," *Facts for Faith*, no. 10 (Q3 2002), 13; Hugh Ross, "Predictive Power: Confirming Cosmic Creation," *Facts for Faith*, no. 9 (Q2 2002), 32–39.

27. Hugh Ross, "The Physics of Sin," *Facts for Faith*, no. 8 (Q1 2002), 46–51; Ross, *Creator and the Cosmos*, 145–167; Hugh Ross, "Physicalism and Free Will," *Facts for Faith*, no. 7 (Q4 2001), 48; Hugh Ross, "Time and the Physics of Sin," in *What God Knows: Time, Eternity, and Divine Knowledge*, ed. Harry Lee Poe and J. Stanley Mattson (Waco, TX: Baylor University Press, 2005), 121–136.

28. Ross, *A Matter of Days*, 97–120, 163–214; Ross, "Time and the Physics of Sin," 46–51.

29. Ross, *Fingerprint of God*, 53–118; Ross, *Creator and the Cosmos*, 23–98, 150–157; Ross, "Predictive Power," 32–39; Hugh Ross, "Facing Up to Big Bang Challenges," *Facts for Faith*, no. 5 (Q1 2001), 42–53.

30. Ross, *Creator and the Cosmos*, 50–53, 150–157; Hugh Ross, "The Haste to Conclude Waste," *Facts & Faith* 11, no. 3 (1997), 1–3; Ross, Samples, and Clark, *Lights in the Sky and Little Green Men*, 33–41, 161–162.

31. Ross, *Fingerprint of God*, 119–138; Ross, *Creator and the Cosmos*, 45–67, 137–212; Hugh Ross, "Anthropic Principle: A Precise Plan for Humanity," *Facts for Faith*, no. 8 (Q1 2002), 24–31; Ross, "Predictive Power," 32–39; Guillermo Gonzalez and Hugh Ross, "Home Alone in the Universe," *First Things*, no. 103 (May 2000), 10–12.

32. Ross, *A Matter of Days*, 218–220; Ross, "Time and the Physics of Sin," 121–136; Hugh Ross, "The Faint Sun Paradox," *Facts for Faith*, no. 10 (Q3 2002), 26–33.

33. Ross, *Creator and the Cosmos*, 56, 179, 183; Ross, "Time and the Physics of Sin," 121–136; Guillermo Gonzalez and Jay W. Richards, *The Privileged Planet: How Our Place in the Cosmos Is Designed for Discovery* (Washington, DC: Regnery, Eagle Publishing, 2004); Ross, "Anthropic Principle," 24–31.

34. Ross, *Creator and the Cosmos*, 56, 179, 183; Ross, "Time and the Physics of Sin," 121–136; Gonzalez and Richards, *Privileged Planet*; Ross, "Anthropic Principle," 24–31.

35. Ross, *Creator and the Cosmos*, 176–178; Rana and Ross, *Origins of Life*, 211–213.

36. Ross, *Creator and the Cosmos*, 184–185; Rana and Ross, *Origins of Life*, 87–88.

37. Rana and Ross, *Origins of Life*, 72–73, 82–85; Hugh Ross, Fazale Rana, and Krista Bontrager, "Magma Ocean," *Creation Update*, Webcast Archives, May 18, 2004, http://www.reasons.org/resources/multimedia/rtbradio/cu_archives/

protected/200401-06archives.shtml, accessed September 21, 2004.

38. Rana and Ross, *Origins of Life*, 63–92.

39. Rana and Ross, *Origins of Life*, 63–92.

40. Rana and Ross, *Origins of Life*, 63–92.

41. Rana and Ross, *Origins of Life*, 93–181; Fazale Rana, "Yet Another Use for 'Junk' DNA," *Facts for Faith*, no. 3 (Q3 2000), 56–57; Fazale R. Rana, "Protein Structures Reveal Even More Evidence for Design," *Facts for Faith*, no. 4 (Q4 2000), 4–5.

42. Rana and Ross, *Origins of Life*, 93–181; Fazale R. Rana, "30% Inefficiency by Design," *Facts for Faith*, no. 6 (Q2 2001), 10–11.

43. Rana and Ross, *Origins of Life*, 74–75, 217; Hugh Ross, "Bacteria Help Prepare Earth for Life," *Connections* 3, no. 1 (2001), 4.

44. Rana and Ross, *Origins of Life*, 219; Hugh Ross, "The Case for Creation Grows Stronger," *Facts & Faith* 4, no. 1 (1990), 1–3.

45. Rana and Ross, *Origins of Life*, 218–220.

46. Rana and Ross, *Origins of Life*, 213–222; Ross, "Faint Sun Paradox," 26–33.

47. Fazale Rana, "New Insight into the Ecology of the Cambrian Fauna: Evidence for Creation Mounts," *Facts for Faith*, no. 3 (Q3 2000), 54–55; Fazale Rana and Hugh Ross, "Exploding with Life!" *Facts for Faith*, no. 2 (Q2 2000), 12–17; Fazale Rana, "The Explosive Appearance of Skeletal Designs," *Facts for Faith*, no. 3 (Q3 2000), 52–53; Hugh Ross, "Biology's Big Bang #2," *Facts & Faith* 7, no. 4 (1990), 3; Fazale R. Rana, "Cambrian Flash," *Connections* 2, no. 1 (2000), 3.

48. Hugh Ross, *The Genesis Question: Scientific Advances and the Accuracy of Genesis*, 2nd ed. (Colorado Springs, CO: NavPress, 2001), 50–53; Rana and Ross, *Origins of Life*, 82–84, 215–231; Hugh Ross, "Creation on the Firing Line," *Facts & Faith* 12, no. 1 (1998), 6–7; Hugh Ross, "Fungus Paints Darker Picture of Permian Catastrophe," *Facts & Faith* 10, no. 2 (1996), 3; Hugh Ross, "Life's Fragility," *Facts & Faith* 8, no. 3 (1994), 4–5; Hugh Ross, "Dinosaurs and Cavemen: The Great Omission?" *Facts & Faith* 6, no. 3 (1992), 6–7; Hugh Ross, "Dinosaurs' Disappearance No Longer a Mystery," *Facts & Faith* 5, no. 3 (1991), 1.

49. Ross, *Genesis Question*, 50–53; Rana and Ross, *Origins of Life*, 82–84, 215–231; Ross, "Creation on the Firing Line," 6–7; Hugh Ross, "Rescued from a Freeze Up," *Facts & Faith* 11, no. 2 (1997), 3; Ross, "Fungus Paints Darker Picture," 3; Ross, "Life's Fragility," 4–5; Ross, "Dinosaurs and Cavemen," 6–7; Ross, "Dinosaurs' Disappearance No Longer a Mystery," 1.

50. Hugh Ross, "The Raising of Lazarus Taxa," *Facts & Faith* 8, no. 3 (1994), 5.

51. Fazale Rana, "Repeatable Evolution or Repeated Creation?" *Facts for Faith*, no. 4 (Q4 2000), 12–21; Fazale Rana, "Convergence: Evidence for a Single Creator," *Facts for Faith*, no. 4 (Q4 2000), 14–20.

52. Rana with Ross, *Who Was Adam?*, 199–225; Ross, *Genesis Question*, 110–115; Fazale R. Rana, "Humans and Chimps Differ," *Connections* 3, no. 3 (2001), 1, 4–5.

53. Ross, *Genesis Question*, 63–65.

54. Hugh Ross, "Petroleum: God's Well-Timed Gift to Mankind," *Connections* 6, no. 3 (2004), 2–3.

55. Ross, "Petroleum," 2–3.

56. Hugh Ross, "Symbiosis—More Complex Than We Knew," *Connections* 1, no. 2 (1999), 2–3.

57. Ross, *A Matter of Days*, 97–109; Fazale Rana, "Extinct Shell Fish Speaks Today," *Connections* 3, no. 2 (2001), 1–2.

58. Ross, *Genesis Question*, 50–54; Fazale R. Rana, "Evolving Robots Challenge Evolution," *Facts for Faith*, no. 5 (Q1 2001), 10–11; Fazale R. Rana, "Marine Body Sizes Add Weight to Creation Model," *Facts for Faith*, no. 5 (Q1 2001), 12–13.
59. Rana, "Extinct Shell Fish Speaks Today," 1–2.
60. Ross, *Creator and the Cosmos*, 139–143; Rana, "30% Inefficiency by Design," 10; Rana, "Yet Another Use for 'Junk' DNA," 56–57; Rana, "Evolving Robots Challenge Evolution," 10–11.
61. Fazale Rana, "New Y Chromosome Studies Continue to Support a Recent Origin and Spread of Humanity," *Facts for Faith*, no. 3 (Q3 2000), 52–53.
62. Fazale Rana, Hugh Ross, and Richard Deem, "Long Life Spans: 'Adam Lived 930 Years and Then He Died,'" *Facts for Faith*, no. 5 (Q1 2001) 18–27; Hugh Ross, "Why Shorter Life Spans?" *Facts for Faith*, no. 5 (Q1 2001), 25; Rana with Ross, *Who Was Adam?*, 111–121.
63. Rana with Ross, *Who Was Adam?*, 77–95; Fazale Rana, "A Fashionable Find," *Connections* 4, no. 1 (2002), 2–3.
64. See chapter 8, "Over-Endowed Humans," pages 159–160.
65. Ross, "Time and the Physics of Sin," 121–136.

Chapter 4: The Biblical Framework of the RTB Creation Model
1. Robert Jastrow, *God and the Astronomers*, 2nd ed. (New York: Norton, 2000). Two by the author relate the story in much more detail. See Hugh Ross, *The Fingerprint of God: Recent Scientific Discoveries Reveal the Unmistakable Identity of the Creator*, 2nd ed. (New Kensington, PA: Whitaker House, 2000), and *The Creator and the Cosmos*, 3rd ed. (Colorado Springs, CO: NavPress, 2001).
2. Jastrow, 107.
3. See, for example, Numbers 23:19; Psalm 119:160; Isaiah 45:18–19; John 8:31–32; 10:35; Titus 1:2; Hebrews 6:18; 1 John 5:6.
4. Psalm 33:11; Lamentations 3:22-24; Malachi 3:6; Hebrews 6:17; James 1:17.
5. Genesis 17:1; 35:11; Job 40:2; Isaiah 44:6; Revelation 4:8; 11:17; 21:22.
6. Proverbs 1–9; John 4:7-12.
7. Article 2, The Belgic Confession in *Ecumenical Creeds and Reformed Confessions* (Grand Rapids, MI: CRC Publications, 1988), 79.
8. Isaiah 55:8-9; Romans 11:34; 1 Corinthians 2:16.
9. 1 Peter 1:12.
10. Hugh Ross, *Beyond the Cosmos: What Recent Discoveries in Astrophysics Reveal About the Glory and Love of God*, 2nd ed. (Colorado Springs, CO: NavPress, 1999), 217–228.
11. Lawrence M. Krauss and Glenn D. Starkman, "Life, the Universe, and Nothing: Life and Death in an Ever-Expanding Universe," *Astrophysical Journal* 531 (2000): 22–30.
12. Carl Sagan in his 1980 PBS television documentary series, *Cosmos*.
13. For a discussion of the biblical content on God's rest day, see Hugh Ross, *A Matter of Days: Resolving a Creation Controversy* (Colorado Springs: CO: NavPress, 2004), 81–84.
14. For a discussion of the biblical content on God's last creation acts on Earth, see Hugh Ross, *The Genesis Question: Scientific Advances and the Accuracy of Genesis*, 2nd ed. (Colorado Springs, CO: NavPress, 2001), 53–57.
15. Jeremiah 33:25.

16. Romans 8:20-22.

17. Ross, *A Matter of Days*, 97–109.

18. For more on the impact of the fall of Adam upon his physical environment, see Ross, *A Matter of Days*, 97–109.

19. Genesis 1:4,10,12,18,21,25.

20. Genesis 1:31.

21. Romans 8:20-23.

22. Genesis 13:16; 15:5; 22:17; Jeremiah 33:22; Hebrews 11:12.

23. Ross, *A Matter of Days*, 90–91.

24. Isaiah 40–48.

25. Francis Brown, S. R. Driver, and Charles A. Briggs, *The Brown-Driver-Briggs Hebrew and English Lexicon* (Peabody, MA: Hendrickson, 1997), 398–401; William Gesenius, *Gesenius' Hebrew and Chaldee Lexicon to the Old Testament Scriptures*, trans. Samuel Prideaux Tregelles (Grand Rapids, MI: Baker, 1979), 341–342; R. Laird Harris, Gleason L. Archer Jr., and Bruce K. Waltke, eds., *Theological Wordbook of the Old Testament* (Chicago: Moody, 1980), 370–371.

26. This implication is explained in Ross, *The Genesis Question*, 27–28.

27. Fazale Rana and Hugh Ross, *Origins of Life: Biblical and Evolutionary Models Face Off* (Colorado Springs, CO: NavPress, 2004), 36–45.

28. Ross, *Genesis Question*, 27–58.

29. 2 Peter 3:10,12.

30. Isaiah 34:4.

31. Charles Seine, "Big Bang's New Rival Debuts with a Splash," *Science* 292 (2001), 189–190.

32. These creatures are identified by the Hebrew noun *nepesh*. Genesis 1 states that God created these species for the first time during the fifth creation day. In one of the creation accounts paralleling Genesis 1, Job 38–42, the strictly physical components of creation are discussed in Job 38 and the *nepesh* creatures in Job 39–41.

33. Only one New Testament passage besides Romans 5:12 pertains to this subject. First Corinthians 15:21-23 states, "For since death came through a man, the resurrection of the dead comes also through a man. For as in Adam all die, so in Christ all will be made alive. But each in his own turn: Christ, the firstfruits; then, when he comes, those who belong to him." This context clearly limits death because of sin to human beings.

34. For more on this topic, see Ross, *A Matter of Days*, 97–109.

35. 1 Corinthians 15:20-23.

36. Ross, *Genesis Question*, 117–187.

37. Michael J. Denton, *Nature's Destiny: How the Laws of Biology Reveal Purpose in the Universe* (New York: Free Press, 1998), 127–132, 251–252, 310–313.

38. Harris, Archer, and Waltke, 213–214.

39. Genesis 1:11-12,21,24-25; Leviticus 19:19.

40. For an in-depth discussion of life beyond the present creation, see Ross, *Beyond the Cosmos*.

41. 1 Corinthians 2:9.

42. Ross, *Creator and the Cosmos*; Hugh Ross, Kenneth Samples, and Mark Clark, *Lights in the Sky and Little Green Men: A Rational Christian Look at UFOs and Extraterrestrials* (Colorado Springs, CO: NavPress, 2002); Rana and Ross, *Origins of Life*; Ross, *A Matter of Days*; Fazale Rana with Hugh Ross, *Who Was Adam? A*

Creation Model Approach to the Origin of Man (Colorado Springs, CO: NavPress, 2005).

Chapter 5: The Cosmos Tests the RTB Creation Model

1. Hugh Ross, *The Creator and the Cosmos: How the Greatest Scientific Discoveries of the Century Reveal God*, 3rd ed. (Colorado Springs, CO: NavPress, 2001); Hugh Ross, *Beyond the Cosmos: What Recent Discoveries in Astrophysics Reveal About the Glory and Love of God*, 2nd ed. (Colorado Springs, CO: NavPress, 1999).

2. D. N. Spergel et al., "Wilkinson Microwave Anisotropy Probe (WMAP) Three Year Results: Implications for Cosmology," *Astrophysical Journal Supplement* (2006), forthcoming. Also available at http://map.gsfc.nasa.gov/m_mm/pub_papers/threeyear.html, accessed May 30, 2006.

3. D. N. Spergel et al., forthcoming; C. L. Bennett et al., "First Year Wilkinson Microwave Anisotropy Probe (WMAP) Observations: Preliminary Maps and Basic Results," *Astrophysical Journal Supplement* 148 (2003): 1–27; J. L. Sievers et al., "Cosmological Parameters from Cosmic Background Imager Observations and Comparisons with BOOMERANG, DASI, and MAXIMA," *Astrophysical Journal* 591 (2003): 599–622; Max Tegmark et al., "Cosmological Parameters from SDSS and WMAP," *Physical Review* D 69 (2004), id: 103501.

4. For an explanation of these theorems and a description of the observational and theoretical evidence in support of these theorems along with citations to the original research, see Ross, *Creator and the Cosmos*, 77–108, 169–174.

5. Ross, *Creator and the Cosmos*, 93–98, 169–174.

6. L. H. Ford and Thomas A. Roman, "Classical Scalar Fields and the Generalized Second Law," *Physical Review* D 64 (2001), id: 024023.

7. Roger Penrose, *Shadows of the Mind: A Search for the Missing Science of Consciousness* (New York: Oxford University Press, 1994), 230.

8. Ross, *Beyond the Cosmos*, 33–35.

9. Genesis 1:1; 2 Timothy 1:9; Titus 1:2.

10. Hebrews 11:3.

11. Matthew 17:1-13; Mark 9:1-13; Luke 9:28-36.

12. John 20:19-31.

13. Ross, *Beyond the Cosmos*, 53–228.

14. Romans 1:18-20.

15. For highly readable descriptions and explanations of these evidences, see James S. Trefil, *The Moment of Creation: Big Bang Physics from Before the First Millisecond to the Present Universe* (New York: Collier Books, 1983), 87–110; John D. Barrow and Joseph Silk, *The Left Hand of Creation: The Origin and Evolution of the Expanding Universe* (New York: Basic Books, 1983), 73–101.

16. J. C. Breckenridge et al., "Macroscopic and Microscopic Entropy of Near-Extremal Spinning Black Holes," *Physics Letters* B 381 (1996): 423–426; Curtis G. Callan Jr. and Juan M. Maldacena, "D-Brane Approach to Black Hole Quantum Mechanics," *Nuclear Physics* B 472 (1996): 591–608; Juan M. Maldacena and Andrew Strominger, "Statistical Entropy of Four-Dimensional Extremal Black Holes," *Physical Review Letters* 77 (1996): 428–429; Andrew Strominger and Cumrun Vafa, "Microscopic Origin of the Bekenstein-Hawking Entropy," *Physics Letters* B 379 (1996): 99–104; Gary Taubes, "How Black Holes May Get String Theory Out of a Bind," *Science* 268 (1995): 1699.

17. Ross, *Beyond the Cosmos*, 40–43.
18. Job 9:8; Psalm 104:2; Isaiah 40:22; 42:5; 44:24; 45:12; 48:13; 51:13; Jeremiah 10:12; 51:15; Zechariah 12:1. For a discussion of exactly what the Bible states about cosmic expansion, see Ross, *Creator and the Cosmos*, 24–26.
19. Job 9:8; Isaiah 44:24; see also Isaiah 45:12.
20. Ross, *Creator and the Cosmos*, 41–42, and references therein.
21. Allan Sandage and Lori M. Lubin, "The Tolman Surface Brightness Test for the Reality of the Expansion. I. Calibration of the Necessary Local Parameters," *Astronomical Journal* 121 (2001): 2271–2288; Lori M. Lubin and Allan Sandage, "The Tolman Surface Brightness Test for the Reality of the Expansion. II. The Effect of the Point-Spread Function and Galaxy Ellipticity on the Derived Photometric Parameters," *Astronomical Journal* 121 (2001): 2289–2300; Lori M. Lubin and Allan Sandage, "The Tolman Surface Brightness Test for the Reality of the Expansion. III. Hubble Space Telescope Profile and Surface Brightness Data for Early-Type Galaxies in Three High-Redshift Clusters," *Astronomical Journal* 122 (2001): 1071–1083; Lori M. Lubin and Allan Sandage, "The Tolman Surface Brightness Test for the Reality of the Expansion. IV. A Measurement of the Tolman Signal and the Luminosity Evolution of Early-Type Galaxies," *Astronomical Journal* 122 (2001): 1084–1103.
22. B. Leibundgut et al., "Time Dilation in the Light Curve of the Distant Type Ia Supernova SN 1995K," *Astrophysical Journal Letters* 466 (1996): L21–L24; A. G. Riess et al., "Time Dilation from Spectral Feature Age Measurements of Type Ia Supernovae," *Astronomical Journal* 114 (1997): 722–729; Garson Goldhaber et al., "Observation of Cosmological Time Dilation Using Type Ia Supernovae as Clocks," in *Thermonuclear Supernovae, Proceedings of the NATO Advanced Study Institute, held in Begur, Girona, Spain, June 20–30, 1995*, NATO Advanced Science Institutes, series C, 486, ed. P. Ruiz-LaPuente, R. Canal, and J. Isern (Dordrecht, Netherlands: Kluwer Academic Publishers, 1997), 777–784; G. Goldhaber et al., "Timescale Stretch Parameterization of Type Ia Supernova B-Band Light Curves," *Astrophysical Journal* 558 (2001): 359–368; Ming Deng and Bradley E. Schaefer, "Time Dilation in the Peak-to-Peak Timescale of Gamma-Ray Bursts," *Astrophysical Journal Letters* 502 (1998): L109–L114.
23. Ross, *Creator and the Cosmos*, 151.
24. Ross, *Creator and the Cosmos*, 150–151.
25. It also exceeds by 10^{97} times the fine-tuning in what probably is humanity's best engineering achievement, a gravity wave telescope that can make length measurements with a precision of one part in 10^{23}.
26. L. Dyson, M. Kleban, and L. Susskind as quoted by Philip Ball, "Is Physics Watching Over Us?" http://www.nature.com/nsu/020812/020812-2.html, accessed August 14, 2002. The preprint to which Philip Ball refers was published in October 2002 by the *Journal of High Energy Physics*: L. Dyson, M. Kleban, and L. Susskind, "Disturbing Implications of a Cosmological Constant," http://ej.iop.org/links/q72/EzjZUjDJyeH0t0iDSa6pPg/jhep102002011.pdf, accessed April 2, 2006. Pertinent quotes are (1) "Some Unknown agent initially started the inflation high up on its potential," 1; (2) "The world started in a state of exceptionally low entropy. . . . However, there is no universally accepted explanation of how the universe got into such a special state," 2; (3) "The question then is whether the origin of the universe can be a naturally occurring fluctuation, or must it be due to an external agent which starts the system out in a specific low entropy state?" 4; (4) "Perhaps the

only reasonable conclusion is that we do not live in a world [universe] with a true cosmological [dark energy] constant," 18.

27. Psalm 104:2.

28. Isaiah 40:22.

29. Stephen W. Hawking, *A Brief History of Time: From the Big Bang to Black Holes* (New York: Bantam Books, 1988), 126–127.

30. Genesis 13:16; 22:17; Jeremiah 33:22; Hebrews 11:12.

31. Robert H. Dicke, "Dirac's Cosmology and Mach's Principle," *Nature* 192 (1961): 440–441.

32. "Fine-Tuning for Life in the Universe," http://www.reasons.org/resources/ apologetics/design_evidences/200412_fine_tuning_for_life_in_the_universe.shtml, accessed September 10, 2005.

33. "Probability for Life on Earth," http://www.reasons.org/resources/apologetics/ design_evidences/200404_probabilities_for_life_on_earth.shtml, accessed September 10, 2005.

34. Freeman J. Dyson, *Disturbing the Universe* (New York: Basic Books, Perseus Books, 1979), 250.

35. Paul Davies, "The Anthropic Principle," *Science Digest* 191, no. 10 (October 1983): 24; Paul Davies, *The Cosmic Blueprint* (New York: Simon & Schuster, 1986), 203.

36. See the lists of these characteristics, "Fine-Tuning for Life on Earth," "Probabilities for Life on Earth," and "Fine-Tuning for Life in the Universe," http://www.reasons .org/resources/apologetics/index.shtml#design_in_the_universe, accessed September 10, 2005.

37. Brandon Carter, "Large Number Coincidences and the Anthropic Principle in Cosmology," *Proceedings of the International Astronomical Union Symposium, No. 63: Confrontation of Cosmological Theories with Observational Data*, ed. M. S. Longair (Dordrecht, Holland and Boston: D. Reidel, 1974), 291–298.

38. Brandon Carter, "The Anthropic Principle and Its Implications for Biological Evolution," *Philosophical Transactions of the Royal Society* A 370 (1983): 347–360.

39. John D. Barrow and Frank J. Tipler, *The Anthropic Cosmological Principle* (New York: Oxford University Press, 1986), 556–570.

40. Barrow and Tipler, 556–570.

41. Ross, *Creator and the Cosmos*, 145–199.

42. Fazale Rana with Hugh Ross, *Who Was Adam? A Creation Model Approach to the Origin of Man* (Colorado Springs, CO: NavPress, 2005), 97–105.

43. Q. R. Ahmad et al., "Measurement of the Rate of $v_e + d \rightarrow p + p + e^-$ Interactions Produced by ^8B Solar Neutrinos at the Sudbury Neutrino Observatory," *Physical Review Letters* 87 (2001): 71301–71305; Hugh Ross and Eric Agol, "Missing Solar Neutrinos Found," *Facts for Faith*, no. 9 (Q2 2002), 11.

44. Adam Eyre-Walker and Peter D. Keightley, "High Genomic Deleterious Mutation Rates in Hominids," *Nature* 397 (1999): 344–347; James F. Crow, "The Odds of Losing at Genetic Roulette," *Nature* 397 (1999): 293–294; Hugh Ross, "Aliens from Another World," *Facts for Faith*, no. 6 (Q2 2001), 30–31.

45. A crude analogy that may help illustrate this point would be the parents of a bride spending long hours and considerable resources preparing for their daughter's wedding that will be celebrated in just a tiny fraction of the preparation time. Because the parents place such a high value and purpose on their daughter and the wedding event, they do not consider the extraordinary investment of time and resources to be a waste.

46. Jeremiah 33:25.
47. John N. Bahcall, Charles L. Steinhardt, and David Schlegel, "Does the Fine-Structure Constant Vary with Cosmological Epoch?" *Astrophysical Journal* 600 (2004): 520–543; P. C. W. Davies, Tamara M. Davis, and Charles H. Lineweaver, "Cosmology: Black Holes Constrain Varying Constants," *Nature* 418 (2002): 602–603; Alexander Y. Potekhin et al., "Testing Cosmological Variability of the Proton-to-Electron Mass Ratio Using the Spectrum of PKS 0528-250," *Astrophysical Journal* 505 (1998): 523–528; D. B. Guenther, "Testing the Constancy of the Gravitational Constant Using Helioseismology," *Astrophysical Journal* 498 (1998): 871–876.
48. E. Peik et al., "Limit on the Present Temporal Variation of the Fine Structure Constant," *Physical Review Letters* 93 (2004), id: 170801.
49. Antoinette Songaila et al., "Measurement of the Microwave Background Temperature at Redshift 1.776," *Nature* 371 (1994): 43–45; David M. Meyer, "A Distant Space Thermometer," *Nature* 371 (1994): 13; K. C. Roth, A. Songaila, L. L. Cowie, and J. Bechtold, "C I Fine-Structure Excitation by the CMBR at z = 1.973," *American Astronomical Society Meeting* 189, no. 122 (December 1996): 17; R. Srianand, P. Petitjean, and C. Leadoux, "The Cosmic Microwave Background Radiation Temperature at Redshift 2.74," *Nature* 408 (2000): 931–935; P. Molaro, S. S. Levshakov, M. Dessauges-Zavadsky, and S. D'Odorico, "The Cosmic Microwave Radiation Temperature at z = 3.025 Toward QSO 0347-3819," *Astronomy and Astrophysics* 381 (2002): L64–L67; E. S. Battistelli et al., "Cosmic Microwave Background Temperature at Galaxy Clusters," *Astrophysical Journal Letters* 580 (2002): L101–L104; D. J. Fixsen and J. C. Mather, "The Spectral Results of the Far-Infrared Absolute Spectrophotometer Instrument on COBE," *Astrophysical Journal* 581 (2002): 817–822; J. C. Mather et al., "Calibrator Design for the COBE Far-Infrared Absolute Spectrophotometer (FIRAS)," *Astrophysical Journal* 512 (1999): 511–520; J. C. Mather et al., "Measurement of the Cosmic Microwave Background Spectrum by the COBE FRAS Instrument," *Astrophysical Journal* 420 (1994): 439–444; Katherine C. Roth, David M. Meyer, and Isabel Hawkins, "Interstellar Cyanogen and the Temperature of the Cosmic Microwave Background Radiation," *Astrophysical Journal Letters* 413 (1993): L67–L71.
50. I give a description of the discovery and subsequent measurements of dark energy and the implication of dark energy on the design and past and present history of the universe in *Creator and the Cosmos*, 45–56.
51. Psalm 19:1-4; 50:6; 97:6; Romans 1:18-20.
52. Bernard E. J. Pagel, *Nucleosynthesis and Chemical Evolution of Galaxies* (New York: Cambridge University Press, 1997), 103–130; Ross, *Creator and the Cosmos*, 57–63.
53. Pagel, 198–320.
54. Joel Baker et al., "Early Planetesimal Melting from an Age of 4.5662 Gyr for Differentiated Meteorites," *Nature* 436 (2005): 1127–1131; C. J. Allègre, G. Manhès, and C. Göpel, "The Age of the Earth," *Geochemica et Cosmochemica Acta* 59 (1995): 1445–1456.

Chapter 6: Planetary Science and Life's Origin Test the RTB Creation Model

1. "Mars Reconnaissance Orbiter Will 'Follow the Water' on Mars," http://www.nasa.gov/mission_pages/MRO/news/mro-feat-083104.html, accessed August 19, 2005.
2. Hugh Ross, *The Creator and the Cosmos: How the Greatest Scientific Discoveries of the Century Reveal God*, 3rd ed. (Colorado Springs, CO: NavPress, 2001); Fazale Rana

and Hugh Ross, *Origins of Life: Biblical and Evolutionary Models Face Off* (Colorado Springs, CO: NavPress, 2004).

3. Michel Mayor and Didier Queloz, "A Jupiter-Mass Companion to a Solar-Type Star," *Nature* 378 (1995): 355–359; Gordon Walker, "Extrasolar Planets: On the Wings of Pegasus," *Nature* 378 (1995): 332–333.
4. A French observatory maintains an up-to-date database on every extrasolar planet that has been discovered at http://www.obspm.fr/planets, accessed May 30, 2006.
5. An orbital resonance between Jupiter and the other gas giant planets would occur if the number of orbits of one or more of the planets Saturn, Uranus, or Neptune about the sun per unit time were a simple fraction of the number of orbits completed by Jupiter during the same period. Such a resonance would cause Jupiter and one or more of the planets Saturn, Uranus, and Neptune to frequently line up and in doing so to exert repeated, strong gravitational tugs on Earth.
6. Guillermo Gonzales, "Spectroscopic Analysis of the Parent Stars of Extrapolar Planetary Systems," *Astronomy and Astrophysics* 334 (1998): 221–238; David T. F. Weldrake et al., "An Absence of Hot Jupiter Planets in 47 Tucanae: Results of a Wide-Field Transit Search," *Astrophysical Journal* 620 (2005): 1043–1051; Guillermo Gonzales, "New Planets Hurt Chances for ETI," *Facts & Faith* 12, no. 4 (1998), 2–4.
7. Robin M. Canup, "Simulations of a Late Lunar-Forming Impact," *Icarus* 168 (2004): 433–456; Herbert Palme, "The Giant Impact Formation of the Moon," *Science* 304 (2004): 977–979.
8. Louis A. Codispoti, "The Limits to Growth," *Nature* 387 (1997): 237; Kenneth H. Coale, "A Massive Phytoplankton Bloom Induced by an Ecosystem-Scale Iron Fertilization Experiment in the Equatorial Pacific Ocean," *Nature* 383 (1996): 495–499.
9. Peter D. Ward and Donald Brownlee, *Rare Earth: Why Complex Life Is Uncommon in the Universe* (New York: Copernicus, Springer-Verlag, 2000), 191–234.
10. William R. Ward, "Comments on the Long-Term Stability of the Earth's Obliquity," *Icarus* 50 (1982): 444–448; Carl D. Murray, "Seasoned Travellers," *Nature* 361 (1993): 586–587; Jacques Laskar and P. Robutel, "The Chaotic Obliquity of the Planets," *Nature* 361 (1993): 608–612; Jacques Laskar, F. Joutel, and P. Robutel, "Stabilization of the Earth's Obliquity by the Moon," *Nature* 361 (1993): 615–617.
11. Dave Waltham, "Anthropic Selection for the Moon's Mass," *Astrobiology* 4 (2004): 460–468.
12. For a detailed explanation with citations to the research literature, see Fazale Rana with Hugh Ross, *Who Was Adam? A Creation Model Approach to the Origin of Man* (Colorado Springs, CO: NavPress, 2005), 102–105.
13. Ross, *Creator and the Cosmos*, 180–199.
14. It is now known that a complex interaction between a giant asteroid-comet belt (the primordial Kuiper Belt) and the planets Jupiter and Saturn caused the Late Heavy Bombardment. Gravitational interactions between the asteroids and comets caused Jupiter and Saturn to drift farther away from the sun. However, because Saturn was closer to the densest part of the belt than Jupiter, Saturn moved outward at a faster rate. When Saturn's orbit reached a point where it made exactly one orbit of the sun for every two orbits of Jupiter, the resulting 1:2 resonance destabilized the entire belt of asteroids and comets. Many of the asteroids and comets were pushed out into the much more distant Oort Cloud. Just as many were sent hurtling into the inner solar system, thereby causing the Late Heavy Bombardment. What made the

bombardment relatively short-lived was that Saturn eventually moved far enough away to break the resonance. (See Richard A. Kerr, "Did Jupiter and Saturn Team Up to Pummel the Inner Solar System?" *Science* 306 (2004): 1676.)

15. Stephen J. Mojzsis, "Lithosphere-Hydrosphere Interactions on the Hadean (>4.0 Ga) Earth," *Astrobiology* 1 (2001): 383; Stephen J. Mojzsis and Graham Ryder, "Accretion to Earth and Moon ~3.85 Ga," in *Accretion of Extraterrestrial Matter Throughout Earth's History,* ed. B. Peuckner-Ehrinbrink and B. Schmitz (New York: Kluwer Academic/Plenum Publishers, 2001); Stephen J. Mojzsis and T. Mark Harrison, "Establishment of a 3.83-Ga Magmatic Age for the Akilia Tonalite (Southern West Greenland)," *Earth and Planetary Science Letters* 202 (2002): 563–576; Ronny Schoenberg et al., "Tungsten Isotope Evidence from ~3.8-Gyr Metamorphased Sediments for Early Meteorite Bombardment of the Earth," *Nature* 418 (2002): 403.

16. A. D. Anbar et al., "Extraterrestrial Iridium, Sediment Accumulation and the Habitability of the Early Earth's Surface," *Journal of Geophysical Research* 106 (2001): 3219–3236; Schoenberg et al., 403–405.

17. David C. Rubie, Christine K. Gessmann, and Daniel J. Frost, "Partitioning of Oxygen During Core Formation on the Earth and Mars," *Nature* 429 (2004): 58–61; Carl B. Agee, "Hot Metal," *Nature* 429 (2004): 33–35.

18. Manfred Schidlowski, "A 3,800-Million-Year Isotopic Record of Life from Carbon in Sedimentary Rocks," *Nature* 333 (1988): 313–318; Manfred Schidlowski, "Carbon Isotopes as Biogeochemical Recorders of Life Over 3.8 Ga of Earth History: Evolution of a Concept," *Precambrian Research* 106 (2001): 117–134; Yuichiro Ueno et al., "Ion Microprobe Analysis of Graphite from Ca. 3.8 Ga Measurements, Isua Supracrustal Belt, West Greenland: Relationship Between Metamorphism and Carbon Isotopic Composition," *Geochimica et Cosmochimica Acta* 66 (2002): 1257–1268; Minik T. Rosing, "^{13}C-Depleted Carbon Microparticles in >3700-Ma Sea-Floor Sedimentary Rocks from West Greenland," *Science* 283 (1999): 674–676; Mink T. Rosing and Robert Frei, "U-Rich Archaean Sea-Floor Sediments from Greenland — Indications of >3700 Ma Oxygenic Photosynthesis," *Earth and Planetary Science Letters* 217 (2004): 237–244.

19. Rosing, 674–676; S. J. Mojzsis et al., "Evidence for Life on Earth Before 3,800 Million Years Ago," *Nature* 384 (1996): 55–59; John M. Hayes, "The Earliest Memories of Life on Earth," *Nature* 384 (1996): 21–22; Manfred Schidlowski, "A 3,800-Million-Year Isotopic Record of Life from Carbon in Sedimentary Rocks," *Nature* 333 (1988): 313–318; Hubert P. Yockey, "Comments on 'Let There Be Life: Thermodynamic Reflections on Biogenesis and Evolution' by Avshalom C. Elitzur," *Journal of Theoretical Biology* 176 (1995): 351; Daniele L. Pinti, Ko Hashizume, and Jun-Ichi Matsuda, "Nitrogen and Argon Signatures in 3.8 to 2.8 Ga Metasediments: Clues on the Chemical State of the Archean Ocean and the Deep Biosphere," *Geochemica et Cosmochimica Acta* 65 (2001): 2309.

20. Rana and Ross, *Origins of Life,* 162–168.

21. L. E. Synder et al., "A Rigorous Attempt to Verify Interstellar Glycine," *Astrophysical Journal* 619 (2005): 914–930.

22. Yi-Jehng Kuan et al., "A Search for Interstellar Pyrimidine," *Monthly Notices of the Royal Astronomical Society* 345 (2003): 650–656.

23. Sandra Pizzarello et al., "The Organic Content of the Tagish Lake Meteorite," *Science* 293 (2001): 229, notes 15 and 28; Jeffrey L. Bada, "A Search for Endogenous Amino Acids in Martian Meteorite ALH84001," *Science* 279 (1998): 362–365;

Keith A. Kvenvolden, "Chirality of Amino Acids in the Murchison Meteorite—A Historical Perspective," in *Book of Abstracts, 12th International Conference on the Origin of Life and the 9th Meeting of the International Society for the Study of the Origin of Life*, July 11–16, 1999, San Diego, California (ISSOL 1999), 41; Daniel P. Glavin et al., "Amino Acids in Martian Meteorite Nakhla," *Book of Abstracts*, ISSOL 1999, 62; Rana and Ross, *Origins of Life*, 95–96, 130–131, 185–190.

24. Juan Oró, "Early Chemical Stages in the Origin of Life," in *Early Life on Earth: Nobel Symposium #84*, ed. Stefan Bengtson (New York: Columbia University Press, 1994), 49–50.

25. Even if a low concentration of a few simple amino acids and nitrogenous bases were one day discovered in some interstellar cloud, such a discovery would prove no boon for naturalistic models for life's origin. All naturalistic models demand that all the biologically required amino acids and nitrogenous bases be available at one location at high concentration levels.

26. Rana and Ross, *Origins of Life*, 63–92.

27. Jon Cohen, "Getting All Turned Around Over the Origin of Life on Earth," *Science* 267 (1995): 1265. In this article Cohen quotes one of the leading origin-of-life researchers, William Bonner, at a conference on the Physical Origin of Homochirality in Life, held in Santa Monica, California, in February 1995, as stating to the assembled scientists, "I spent 25 years looking for terrestrial mechanisms for homochirality and trying to investigate them and didn't find any supporting evidence. Terrestrial explanations are impotent or nonviable."

28. Robert M. Hazen, "Life's Rocky Start," *Scientific American* (April 2001): 77–85.

29. G. Balavoine, A. Monadpour, and H. B. Kagan, "Preparation of Chiral Compounds with High Optical Purity by Irradiation with Circularly Polarized Light: A Model Reaction for the Prebiotic Generation of Optical Activity," *Journal of the American Chemical Society* 96 (1974): 5152–5158.

30. Jose J. Flores, William A. Bonner, and Gail A. Massey, "Asymmetric Photolysis of (RS)-Leucine with Circularly Polarized Ultraviolet Light," *Journal of the American Chemical Society* 99 (1977): 3622–3625.

31. Mark M. McKinnon, "Statistical Modeling of the Circular Polarization in Pulsar Radio Emission and Detection Statistics of Radio Polarimetry," *Astrophysical Journal* 568 (2002): 302–311.

32. Yoshinori Takano et al., "Asymmetric Photolysis of (DL)-Isovaline by Synchrotron Radiation," in *Book of Abstracts*, 13th International Conference on the Origin of Life and the 10th Meeting of the International Society for the Study of the Origin of Life (ISSOL 2002), June 30–July 5, 2002, Oaxaca, Mexico, 92–93. The figure, 1.12 percent, was presented in a poster paper.

33. Werner Kuhn, "The Physical Significance of Optical Rotary Power," *Transactions of the Faraday Society* 26 (1930): 293–308; E. U. Condon, "Theories of Optical Rotary Power," *Reviews of Modern Physics* 9 (1937): 432–457.

34. Rana and Ross, *Origins of Life*, 95–101, 123–133, 171–222.

35. F. H. C. Crick and Leslie E. Orgel, "Directed Panspermia," *Icarus* 19 (1973): 341–346; Francis Crick, *Life Itself: Its Origin and Nature* (New York: Simon & Schuster, 1981).

36. Hugh Ross, Kenneth Samples, and Mark Clark, *Lights in the Sky and Little Green Men: A Rational Christian Look at UFOs and Extraterrestrials* (Colorado Springs, CO: NavPress, 2002), 55–64.

37. Genesis 1:1; John 1:1-3; Colossians 1:15-17; 2 Timothy 1:9; Titus 1:2; Hebrews 11:3.
38. Jeremiah 23:24.
39. Stuart Kauffman, *Investigations* (New York: Oxford University Press, 2000), 35, 43, 46, 151; Hubert P. Yockey, *Information Theory and Molecular Biology* (New York: Cambridge University Press, 1992), 289; Niels Bohr, "Light and Life," *Nature* 131 (1933): 421–423, 457–459.
40. John C. Armstrong, Llyd E. Wells, and Guillermo Gonzalez, "Rummaging Through Earth's Attic for Remains of Ancient Life," *Icarus* 160 (2002), 183–196.
41. Fazale Rana and Hugh Ross, *Origins of Life.*
42. Armstrong, Wells, and Gonzalez, 183–196; S. A. Finney, W. B. Tonks, and H. J. Melosh, "Statistical Evolution of Impact Ejecta from the Earth: Implications for Transfer to Other Solar System Bodies," *Contribution #698, Twentieth Lunar and Planetary Science Conference*, March 13–17, 1989, Houston, Texas.
43. Stephen J. Mojzsis, T. Mark Harrison, and Robert T. Pidgeon, "Oxygen-Isotope Evidence from Ancient Zircons for Liquid Water at the Earth's Surface—4,300 Myr Ago," *Nature* 409 (2001): 178–181.

Chapter 7: Life's History on Earth Tests the RTB Creation Model

1. *The Blind Watchmaker* and *Climbing Mount Improbable* are the titles of two books promoting a naturalistic explanation for life's history on Earth by Britain's famous zoologist and anticreationist, Richard Dawkins.
2. Hugh Ross, *The Genesis Question: Scientific Advances and the Accuracy of Genesis*, 2nd ed. (Colorado Springs, CO: NavPress, 2001).
3. Arcady R. Mushegian and Eugene V. Koonin, "A Minimum Gene Set for Cellular Life Derived by Comparison of Complete Bacterial Genomes," *Proceedings of the National Academy of Sciences, USA* 93 (1996): 10268–10273; Nikos Kyrpides et al., "Universal Protein Families and the Functional Contents of the Last Universal Common Ancestor," *Journal of Molecular Evolution* 49 (1999): 413–423; Jack Manloff, "The Minimal Cell Genome: On Being the Right Size," *Proceedings of the National Academy of Sciences, USA* 93 (1996): 10004–10006; Mitsuhiro Itaya, "An Estimation of Minimal Genome Size Required for Life," *FEBS Letters* 362 (1995): 257–260; Clyde A. Hutchinson III et al., "Global Transposon Mutagenesis and a Minimal Mycoplasma Genome," *Science* 286 (1999): 2165–2169; Brian J. Ackerley et al., "A Genome-Scale Analysis for Identification of Genes Required for Growth or Survival of *Haemophilus Influenzae*," *Proceedings of the National Academy of Sciences, USA* 99 (2002): 966–971; Rosario Gil et al., "Extreme Genome Reduction in *Buchnera* spp: Toward the Minimal Genome Needed for Symbiotic Life," *Proceedings of the National Academy of Sciences, USA* 99 (2002): 4454–4458.
4. Minik T. Rosing and Robert Frei, "U-Rich Archaen Sea-Floor Sediments from Greenland—Indications of >3700 Ma Oxygenic Photosynthesis," *Earth and Planetary Science Letters* 6907 (2003): 1–8.
5. Don Cowan, "Use Your Neighbour's Genes," *Nature* 407 (2000): 466–467; Andreas Ruepp et al., "The Genome Sequence of the Thermoacidophilic Scavenger *Thermoplasma acidophilum*," *Nature* 407 (2000): 508–513; Gerard Deckert et al., "The Complete Genome of the Hyperthermophilic Bacterium *Aquifex aeolicus*," *Nature* 392 (1998): 353–358; Alexei I. Slesarev et al., "The Complete Genome of Hyperthermophile *Methanopyrus kandleri* AV19 and Monophyly of Archael Methanogens," *Proceedings of the National Academy of Sciences, USA* 99 (2002):

4644–4469; Virginia Morell, "Life's Last Domain," *Science* 273 (1996): 1043–1045;
Carol J. Bult et al., "Complete Genome Sequence of the Methanogenic Archaeon,
Methanococcus jannaschii," *Science* 273 (1996): 1058–1073; Elizabeth Pennisi,
"Microbial Genomes Come Tumbling In," *Science* 277 (1997): 1433; Karen E.
Nelson et al., "Evidence for Lateral Gene Transfer Between Archaea and Bacteria
from Genome Sequence of *Thermotoga maritime*," *Nature* 399 (1999): 323–329;
Colin Patterson, *Evolution*, 2nd ed. (Ithaca, NY: Comstock, 1999), 23.

6. Crisogono Vasconcelos and Judith A. McKenzie, "Sulfate Reducers—Dominant
   Players in a Low-Oxygen World?" *Science* 290 (2000): 1711–1712.

7. John Emsley, *The Elements*, 3rd ed. (Oxford, UK: Clarendon Press, 1998), 24, 40, 56,
   58, 60, 62, 78, 102, 106, 122, 130, 138, 152, 160, 188, 198, 214, 222, 230.

8. Matthias Labrenz et al., "Formation of Sphalerite (ZnS) Deposits in Natural Biofilms
   of Sulfate-Reducing Bacteria," *Science* 290 (2000): 1744–1747.

9. Tyler Volk and David Schwartzman, "Biotic Enhancement of Weathering and
   the Habitability of Earth," *Nature* 340 (1989): 457–460; Richard Monastersky,
   "Supersoil," *Science News* 136 (1989): 376–377.

10. I.-Juliana Sackmann and Arnold I. Boothroyd, "Our Sun. V. A Bright Young Sun
    Consistent with Helioseismology and Warm Temperatures on Ancient Earth and
    Mars," *Astrophysical Journal* 583 (2003): 1024–1039.

11. Sackmann and Boothroyd, 1024–1039.

12. Hugh Ross, *The Creator and the Cosmos: How the Greatest Scientific Discoveries of the
    Century Reveal God*, 3rd ed. (Colorado Springs, CO: NavPress, 2001), 180–181.

13. Jihad Touma and Jack Wisdom, "Nonlinear Core-Mantle Coupling," *Astronomical
    Journal* 122 (2001): 1030–1050; Gerald Schubert and Keke Zhang, "Effects
    of an Electrically Conducting Inner Core on Planetary and Stellar Dynamos,"
    Astrophysical Journal 557 (2001): 930–942; M. H. Acuna et al., "Magnetic Field and
    Plasma Observations at Mars: Initial Results of the Mars Global Surveyor Mission,"
    Science 279 (1998): 1676–1680; Peter Olson, "Probing Earth's Dynamo," *Nature* 389
    (1997): 337; Weiji Kuang and Jeremy Bloxham, "An Earth-Like Numerical Dynamo
    Model," *Nature* 389 (1997): 371–374; Xiaodong Song and Paul G. Richards,
    "Seismological Evidence for Differential Rotation of the Earth's Inner Core," *Nature*
    382 (1997): 221–224; Wei-jia Su, Adam M. Dziewonski, and Raymond Jeanloz,
    "Planet Within a Planet: Rotation of the Inner Core of the Earth," *Science* 274
    (1996): 1883–1887.

14. Stephen H. Kirby, "Taking the Temperature of Slabs," *Nature* 403 (2000): 31–34.

15. Peter D. Ward and Donald Brownlee, *Rare Earth: Why Complex Life Is Uncommon in
    the Universe* (New York: Copernicus, 2000), 191–234.

16. Donald E. Canfield and Andreas Teske, "Late Proterozoic Rise in Atmospheric
    Oxygen Concentration Inferred from Phylogenetic and Sulfur-Isotope Studies,"
    Nature 382 (1996): 127–132; Donald E. Canfield, "A New Model for Proterozoic
    Ocean Chemistry," *Nature* 396 (1998): 450–453; John M. Hayes, "A Lowdown on
    Oxygen," *Nature* 417 (2002): 127.

17. Paul G. Falkowski et al., "The Rise of Oxygen over the Past 205 Million Years and
    the Evolution of Large Placental Mammals," *Science* 309 (2005): 2202–2204.

18. J. S. Seewald, "Organic-Inorganic Interactions in Petroleum-Producing Sedimentary
    Basins," *Nature* 426 (2003): 327–333.

19. I. M. Head, D. M. Jones, and S. R. Larter, "Biological Activity in the Deep
    Subsurface and the Origin of Heavy Oil," *Nature* 426 (2003): 344–352.

20. N. White, M. Thompson, and T. Barwise, "Understanding the Thermal Evolution of Deep-Water Continental Margins," *Nature* 426 (2003): 334–343.
21. A recent paper in *Science* identifies body size as an important extinction factor: Marcel Cardillo et al., "Multiple Causes of High Extinction Risk in Large Mammal Species," *Science* 309 (2005): 1239–1241.
22. Ecologists Paul and Anne Ehrlich in their book *Extinction* (New York: Ballantine Books, 1981), 23, wrote, "The production of a new animal species in nature has yet to be documented. Biologists have not been able to observe the entire sequence of one species being transformed into two or more. . . . Biologists have been able to observe innumerable examples of animal and plant species that appear to be in various stages of splitting. But, in the vast majority of cases, the rate of change is so slow that it has not even been possible to detect an increase in the amount of differentiation over the decades that have been available for observation."
23. Carl Zimmer, *At the Water's Edge: Macroevolution and the Transformation of Life* (New York: The Free Press, 1998); Michael Behe, David Berlinkski, William F. Buckley Jr., Phillip Johnson, Barry Lynn, Kenneth Miller, Michael Ruse, and Eugenie Scott, "PBS Debate on Creation and Evolution," *Firing Line,* December 19, 1997.
24. Mandyam V. Srinivasan, "When One Eye Is Better Than Two," *Nature* 399 (1999): 305–307; J. D. Pettigrew and S. P. Collin, "Terrestrial Optics in an Aquatic Eye: The Sandlance, *Limnichthyes fasciatus* (Creediidae Teleostei)," *Journal of Comparative Physiology* A 177 (1995): 397–408; John D. Pettigrew, Shaun P. Collin, and Matthias Ott, "Convergence of Specialised Behaviour, Eye Movements and Visual Optics in the Sandlance (Teleostei) and the Chameleon (Reptilia)," *Current Biology* 9 (1999): 421–424.
25. Katherine L. Moulton and Robert A. Berner, "Quantification of the Effect of Plants on Weathering: Studies in Iceland," *Geology* 26 (October 1998): 895–898.
26. Neville J. Woolf, "What Is an Earth-Like Planet?" Abstract #926, Abstracts of the Biennial Meeting of the NASA Astrobiology Institute, April 10–14, 2005, *Astrobiology* 5 (2005): 186–187.
27. Isabelle Basile-Doelsch, Jean Dominique Meunier, and Claude Parron, "Another Continental Pool in the Terrestrial Silicon Cycle," *Nature* 433 (2005): 399–402; Philip W. Boyd et al., "The Decline and Fate of an Iron-Induced Subarctic Phytoplankton Bloom," *Nature* 428 (2004): 549–553.

Chapter 8: The Origin and History of Humanity Test the RTB Creation Model

1. *The War of the Worlds: Mars' Invasion of Earth, Inciting Panic and Inspiring Terror from H. G. Wells to Orson Welles and Beyond* (Naperville, IL: Sourcebooks MediaFusion, 2005), 80.
2. *War of the Worlds,* 81.
3. *War of the Worlds,* 81.
4. *War of the Worlds,* 83.
5. "Humans on Display at London's Zoo," *CBS News,* August 26, 2005, http://www.cbsnews.com/stories/2005/08/26/world/main798423.shtml, accessed August 29, 2005. See also "Crowds Go Ape over 'Humans' Zoo Exhibit," MSNBC, August 26, 2005, http://www.msnbc.msn.com/id/9087023/, accessed August 29, 2005.
6. Fazale Rana with Hugh Ross, *Who Was Adam? A Creation Model Approach to the Origin of Man* (Colorado Springs, CO: NavPress, 2004).

7. Quoted by Frank J. Tipler in "Intelligent Life in Cosmology," *International Journal of Astrobiology* 2 (2003): 142.
8. Brandon Carter, "The Anthropic Principle and Its Implications for Biological Evolution," *Philosophical Transactions of the Royal Society* A 370 (1983): 347–360; John D. Barrow and Frank J. Tipler, *The Anthropic Cosmological Principle* (New York: Oxford University Press, 1986), 510–573.
9. Matthias Krings et al., "Neanderthal DNA Sequences and the Origin of Modern Humans," *Cell* 90 (1997): 10–30; Ryk Ward and Chris Stringer, "A Molecular Handle on the Neanderthals," *Nature* 388 (1997): 37; Patricia Kahn and Ann Gibbons, "DNA from an Extinct Human," *Science* 277 (1997): 176–178; Matthias Krings et al., "DNA Sequence of the Mitochondrial Hypervariable Region II from the Neanderthal Type Specimen," *Proceedings of the National Academy of Sciences, USA* 96 (1999): 5581–5585; Matthias Höss, "Neanderthal Population Genetics," *Nature* 404 (2000): 453–454; Igor V. Ovchinnikov et al., "Molecular Analysis of Neanderthal DNA from the Northern Caucasus," *Nature* 404 (2000): 490–493; Matthias Krings et al., "A View of Neanderthal Genetic Diversity," *Nature Genetics* 26 (2000):144–146; Ralf W. Schmitz et al., "The Neanderthal Type Site Revisited: Interdisciplinary Investigations of Skeletal Remains from the Neander Valley, Germany," *Proceedings of the National Academy of Sciences, USA* 99 (2002): 13342–13347; David Serre et al., "No Evidence of Neanderthal mDNA Contribution to Early Modern Humans," *PLoS Biology* 2, no. 3 (2004): e57; M. Currat and L. Excoffier, "Modern Humans Did Not Admix with Neanderthals During Their Range Expansions into Europe," *PLoS Biology* 2, no. 12 (2004): e421; Cedric Beauval et al., "A Late Neanderthal Femur from Les Rochers-de-Villeneuve, France," *Proceedings of the National Academy of Sciences* 102 (2005): 7085–7090.
10. David Caramelli et al., "Evidence for a Genetic Discontinuity Between Neanderthals and 24,000-Year-Old Anatomically Modern Europeans," *Proceedings of the National Academy of Sciences, USA* 100 (2003): 6593–6597; Oliva Handt et al., "Molecular Genetic Analysis of the Tyrolean Ice Man," *Science* 264 (1994): 1775–1778; Giulietta Di Benedetto et al., "Mitochondrial DNA Sequences in Prehistoric Human Remains from the Alps," *European Journal of Human Genetics* 8 (2000): 669–677.
11. H. Q. Coqueugniot et al., "Early Brain Growth in *Homo erectus* and Implications for Cognitive Ability," *Nature* 431 (2004), 299–302.
12. Marcia S. Ponce de Leon and Christopher P. E. Zollikofer, "Neanderthal Cranial Ontogeny and Its Implications for Late Hominid Diversity," *Nature* 412 (2001): 534–538; B. Bower, "Neanderthals, Humans May Have Grown Apart," *Science News* 160 (2001): 71; Fernando V. Ramirez Rozzi and José Maria Bermudez de Castro, "Surprisingly Rapid Growth in Neanderthals," *Nature* 428 (2004): 936–939; Jay Kelley, "Neanderthal Teeth Lined Up," *Nature* 428 (2004): 904–905.
13. Linda Vigilant et al., "African Populations and the Evolution of Human Mitochondrial DNA," *Science* 253 (1991): 1503–1507; Margellen Ruvolo et al., "Mitochondrial COII Sequence and Modern Human Origins," *Molecular Biology and Evolution* 10 (1993): 1115–1135; Stephen T. Sherry et al., "Mismatch Distributions of mDNA Reveal Recent Human Population Expansions," *Human Biology* 66 (1994): 761–775; Satoshi Horai et al., "Recent African Origin of Modern Humans Revealed by Complete Sequences of Hominid Mitochondrial DNAs," *Proceedings of the National Academy of Sciences, USA* 92 (1995): 532–536; M. Hasegawa and S. Horai, "Time of the Deepest Root for Polymorphism in Human Mitochondrial

DNA," *Journal of Molecular Evolution* 32 (1991): 37–42; Mark Stoneking et al., "New Approaches to Dating Suggest a Recent Age for the Human mtDNA Ancestor," *Philosophical Transactions of the Royal Society of London* B 337 (1992): 167–175; Max Ingman et al., "Mitochondrial Genome Variation and the Origin of Modern Humans," *Nature* 408 (2000): 708–713; S. Blair Hedges, "A Start for Population Genomics," *Nature* 408 (2000): 652–653; Ann Gibbons, "Calibrating the Mitochondrial Clock," *Science* 279 (1998): 28–29. For a thorough review of the genetic origins of human beings, see Rana and Ross, *Who Was Adam?*

14. L. Simon Whitfield et al., "Sequence Variation of the Human Y Chromosome," *Nature* 378 (1995): 379–380; Jonathan K. Pritchard et al., "Population Growth of Human Y Chromosomes: A Study of Y Chromosome Microsatellites," *Molecular Biology and Evolution* 16 (1999): 1791–1798; Russell Thomson et al., "Recent Common Ancestry of Human Y Chromosomes: Evidence from DNA Sequence Data," *Proceedings of the National Academy of Sciences, USA* 97 (2000): 7360–7365; Peter A. Underhill et al., "Y Chromosome Sequence Variation and the History of Human Populations," *Nature Genetics* 26 (2000): 358–361; Ann Gibbons, "Y Chromosome Shows That Adam Was an African," *Science* 278 (1997): 804–805; Mark Seielstad et al., "A View of Modern Human Origins from Microsatellite Variation," *Genome Research* 9 (1999): 558–567; Ornelia Semino et al., "Ethiopians and Khoisan Share the Deepest Clades of the Human Y Chromosome Phylogeny," *American Journal of Human Genetics* 70 (2002): 265–268; For a thorough review of the genetic origins of human beings, see Rana and Ross, *Who Was Adam?*

15. Genesis 10:1–11:9; Vincent Macaulay et al., "Single, Rapid Coastal Settlement of Asia Revealed by Analysis of Complete Mitochondrial Genomes," *Science* 308 (2005): 1034–1036.

16. Tatsuya Anzai et al., "Comparative Sequencing of Human and Chimpanzee MHC Class 1 Regions Unveils Insertions/Deletions as the Major Path to Genome Divergence," *Proceedings of the National Academy of Sciences, USA* 100 (2003): 7708–7713; J. W. Thomas et al., "Comparative Analysis of Multi-Species Sequences from Targeted Genomic Regions," *Nature* 424 (2003): 788–793; Ulfur Arnason et al., "Comparison Between the Complete Mitochondrial DNA Sequences of Homo and the Common Chimpanzee Based on Nonchimeric Sequences," *Journal of Molecular Evolution* 42 (1996): 145–152; The International Chimpanzee Chromosome 22 Consortium, "DNA Sequence and Comparative Analysis of Chimpanzee Chromosome 22," *Nature* 429 (2004): 382–388; Jean Weisenbach, "Genome Sequencing Differences with the Relatives," *Nature* 429 (2004): 353–355.

17. Dennis Normile, "Gene Expression Differs in Human and Chimp Brains," *Science* 292 (2001): 44–45; Mario Caceres et al., "Elevated Gene Expression Levels Distinguish Human from Non-Human Primate Brains," *Proceedings of the National Academy of Sciences, USA* 100 (2003): 13030–13035; Monica Uddin et al., "Sister Grouping of Chimpanzees and Humans as Revealed by Genome-Wide Phylogenic Analysis of Brain Gene Expression Profiles," *Proceedings of the National Academy of Sciences, USA* 101 (2004): 2957–2962; Philipp Khaitovich et al., "Regional Patterns of Gene Expression in Human and Chimpanzee Brains," *Genome Research* 14 (2004): 1462–1473; The International Chimpanzee Chromosome 22 Consortium, 382–388; Weisenbach, 353–355; Peter A. Jones and Daiya Takai, "The Role of DNA Methylation in Mammalian Epigenetics," *Science* 293 (2001): 1068–1070; Todd M. Preuss et al., "Human Brain Evolution Insights from Microarrays," *Nature Genetics Reviews* 5 (2004): 850–860.

18. Genesis 1:28-30.
19. Genesis 3:17; 9:2.
20. Gary K. Meffe, C. Ronald Carroll, and contributors, *Principles of Conservation Biology*, 2nd ed. (Sunderland, MA: Sinauer Associates, 1997), 87–156; John Alroy, "A Multispecies Overkill Simulation of the End-Pleistocene Megafaunal Mass Extinction," *Science* 292 (2001): 1893–1896; Richard G. Roberts et al., "New Ages for the Last Australian Megafauna: Continent-Wide Extinction About 46,000 Years Ago," *Science* 292 (2001): 1888–1892; Paul R. and Anne H. Ehrlich, *Extinction* (New York: Ballantine Books, 1981), 20–21; Jeffrey K. McKee et al., "Forecasting Global Biodiversity Threats Associated with Human Population Growth," *Biological Conservation* 115 (2003): 161–164; Leigh Dayton, "Mass Extinctions Pinned on Ice Age Hunters," *Science* 292 (2001): 1819; Gerardo Ceballos and Paul R. Ehrlich, "Mammal Population Losses and the Extinction Crisis," *Science* 296 (2002): 904–907; David W. Steadman, "Prehistoric Extinctions of Pacific Island Birds: Biodiversity Meets Zooarchaeology," *Science* 267 (1995): 1123–1130; "Human Impact on the Earth: Journey into New Worlds," http://www.sacredbalance.com/web/drilldown.html?sku=35, accessed June 29, 2005; The Savory Center, "The Late Pleistocene Extinctions," http://www.holisticmanagement.org/oll_late.cfm?cfid=834 659&cftoken=58449918, accessed June 29, 2005.
21. Christopher Stringer and Robin McKie, *African Exodus: The Origins of Humanity* (New York: Holt, 1997), 165–166; P. S. Martin and R. G. Klein, eds., *Quaternary Extinctions: A Prehistoric Revolution* (Tuscon, AZ: Arizona University Press, 1984).
22. Pritchard et al., 1791–1798; Thomson et al., 7360–7365; Underhill et al., 358–361; Whitfield et al., 379–380.
23. Ingman et al., 708–713; Hedges, 652–653.
24. Lois A. Tully et al., "A Sensitive Denaturing Gradient-Gel Electrophoresis Assay Reveals a High Frequency of Heteroplasmy in Hypervariable Region 1 of the Human mtDNA Central Region," *American Journal of Human Genetics* 67 (2000): 432–443; Gibbons, 28–29.
25. Gibbons, 28–29; Hugh Ross and Sam Conner, "Eve's Secret to Growing Younger," *Facts & Faith* 12, no. 1 (1998), 1–2.
26. Rana and Ross, *Who Was Adam?* 46–47.
27. Scott A. Elias et al., "Life and Times of the Bering Land Bridge," *Nature* 382 (1996): 61–63; Heiner Josenhans et al., "Early Humans and Rapidly Changing Holocene Sea Levels in the Queen Charlotte Islands—Hecate Strait, British Columbia, Canada," *Science* 277 (1997): 71.
28. Richard G. Klein with Blake Edgar, *The Dawn of Human Culture: A Bold New Theory on What Sparked the "Big Bang" of Human Consciousness* (New York: Wiley, 2002), 230–237.
29. Richard G. Klein, *The Human Career: Human Biological and Cultural Origins*, 2nd ed. (Chicago: University of Chicago Press, 1999), 520–529; Olga Soffer, "Late Paleolithic," *Encyclopedia of Human Evolution and Prehistory*, 2nd ed., ed. Eric Delson et al. (New York: Garland, 2000), 375–380; Alison S. Brooks, "Later Stone Age," *Encyclopedia of Human Evolution and Prehistory*, 2nd ed., 380–382.
30. Ralf Kittler, Manfred Kayser, and Mark Stoneking, "Molecular Evolution of *Pediculus humanus* and the Origin of Clothing," *Current Biology* 13 (2003): 1414–1417.
31. Soffer, 375–380; Klein with Edgar, 11–15; Klein, 512–515; Steven L. Kuhn et al., "Ornaments of the Earliest Upper Paleolithic: New Insights from the Levant,"

Proceedings of the National Academy of Sciences, USA 98 (2001): 7641–7646.

32. Roger Lewin, *Principles of Human Evolution: A Core Textbook* (Malden, MA: Blackwell Science, 1998), 469–474; Rex Dalton, "Lion Man Takes Pride of Place as Oldest Statue," http://www.nature.com/news/2003/030901/full/030901-6.html, accessed June 25, 2005; Nicholas J. Conrad, "Paleolithic Ivory Sculptures from Southwestern Germany and the Origins of Figurative Art," *Nature* 426 (2003): 830–832; Soffer, 375–380; Achim Schneider, "Ice-Age Musicians Fashioned Ivory Flute," http://nature.com/news/2004/041213/pf/041213-14_pf.html, accessed June 25, 2005; Tim Appenzeller, "Evolution or Revolution?" *Science* 283 (1999): 1451–1454; Klein, 550–553.

33. Nations with per capita income exceeding $20,000 all have birth rates less than the replacement rate. Interestingly, the availability of superior birth control methods does not play the most significant role in lowering the birth rate. That distinction belongs to the availability of electric lights and the degree of urbanization. Also, the age at which men and women have their first child rises in direct proportion to the level of technology that they enjoy. See Larry Weiser, Bob Enright, and George Langelett, "World Population Change: Boom or Bust?" http://www.uwsp.edu/business/economicswisconsin/e_lecture/pop_sum.htm, accessed April 10, 2006; CIA World Factbook, "Population Growth Rate (%) 2005," http://www.photius.com/rankings/population/population_growth_rate_2005_1.html, accessed April 10, 2006; CIA World Factbook, "GDP Per Capita 2005," http://www.photius.com/rankings/economy/gdp_per_capita_2005_0.html, accessed April 10, 2006.

Chapter 9: The Why Challenge

1. "Creation/Evolution: The Eternal Debate," http://crevo.bestmessageboard.com, accessed December 30, 2004.
2. Fazale Rana with Hugh Ross, *Who Was Adam? A Creation Model Approach to the Origin of Man* (Colorado Springs, CO: NavPress, 2005); Hugh Ross, *A Matter of Days: Resolving a Creation Controversy* (Colorado Springs, CO: NavPress, 2004); Fazale Rana and Hugh Ross, *Origins of Life: Biblical and Evolutionary Models Face Off* (Colorado Springs, CO: NavPress, 2004); Hugh Ross, *The Creator and the Cosmos: How the Greatest Scientific Discoveries of the Century Reveal God*, 3rd ed. (Colorado Springs, CO: NavPress, 2001); Hugh Ross, *The Genesis Question: Scientific Advances and the Accuracy of Genesis*, 2nd ed. (Colorado Springs, CO: NavPress, 2001); Hugh Ross, Kenneth Samples, and Mark Clark, *Lights in the Sky and Little Green Men: A Rational Christian Look at UFOs and Extraterrestrials* (Colorado Springs, CO: NavPress, 2002).
3. For a thorough discussion of what the Bible teaches about death, decay, work, pain, and suffering before the advent of humanity, see Ross, *A Matter of Days*, 97–120.
4. Job 38:34–39:40; Psalms 104:10-28; 145:7-16; 147:8-18.
5. Melissa Wray, "Overpopulation of Elephants: A Mighty Dilemma," *Africa's Online Magazine*, http://www.scienceinafrica.co.za/2004/september/elephant.htm, accessed August 26, 2005.
6. Greg Dwyer, Jonathon Dushoff, and Susan Harrell Yee, "The Combined Effects of Pathogens and Predators on Insect Outbreaks," *Nature* 430 (2004): 341–345; Lewi Stone, "A Three-Player Solution," *Nature* 430 (2004): 299–300.
7. Harvey Lodish et al., *Molecular Cell Biology*, 4th ed. (New York: Freeman, 2000), 297–303; Wen-Hsiung Li, *Molecular Evolution* (Sunderland, MA: Sinauer Associates, 1998), 395–399.

8. Edward E. Max, "Plagiarized Errors and Molecular Genetics: Another Argument in the Evolution-Creation Controversy," http://www.talkorigins.org/faqs/molgen/, accessed May 9, 2005.
9. Max, http://www.talkorigins.org/faqs/molgen/; Lodish et al., 299–301, 303.
10. R. N. Mantegna et al., "Linguistic Features of Noncoding DNA Sequences," *Physical Review Letters* 73 (1994): 3169–3172.
11. Stephen Jay Gould, *The Panda's Thumb: More Reflections in Natural History* (New York: Norton, 1980).
12. Peter Gordon, "The Panda's Thumb Revisited: An Analysis of Two Arguments Against Design," *Origins Research*, no. 7 (1984): 12–14.
13. Hideki Endo et al., "Role of the Giant Panda's 'Pseudo-Thumb,'" *Nature* 397 (1999): 309.
14. Endo et al., 310.
15. Amélie Davis and Xiao-Hai Yan, "Hurricane Forcing on Chlorophyll-a Concentration Off the Northeast Coast of the U.S.," *Geophysical Research Letters* 31 (2004): L17304, doi:10.1029/2004GL020668.
16. D. M. Murphy et al., "Influence of Sea-Salt on Aerosol Radiative Properties in the Southern Ocean Marine Boundary Layer," *Nature* 392 (1998): 62–65.
17. Nicholas R. Bates, Anthony H. Knap, and Anthony F. Michaels, "Contribution of Hurricanes to Local and Global Estimates of Air-Sea Exchange of CO_2," *Nature* 395 (1998): 58–61.
18. Peter D. Moore, "Fire Damage Soils Our Forest," *Nature* 384 (1996): 312–313.
19. A. U. Mallik, C. H. Gimingham, and A. A. Rahman, "Ecological Effects of Heather Burning I. Water Infiltration, Moisture Retention, and Porosity of Surface Soil," *Journal of Ecology* 72 (1984): 767–776.

Chapter 10: The Power of Proof

1. *The War of the Worlds: Mars' Invasion of Earth, Inciting Panic and Inspiring Terror from H. G. Wells to Orson Welles and Beyond* (Naperville, IL: Sourcebooks MediaFusion, 2005), 43.
2. This exchange of questions and responses took place after my lecture on the scientific evidences for the existence of God at an Atheists United meeting on the Cypress College campus in Southern California in 1988. After a very long question-and-answer session in which over a sixth of 300 plus attendees came to the microphone to challenge me with their questions, I then posed these two questions to the audience.
3. I asked this question and a few follow-up questions after I spent an afternoon and part of an evening answering questions from an assembled gathering in 1999 in Albuquerque, New Mexico, of about 35 young-earth creationist apologists.
4. I. S. Shklovskii and Carl Sagan, *Intelligent Life in the Universe* (San Francisco: Holden-Day, 1966).
5. Hugh Ross, *The Creator and the Cosmos: How the Greatest Scientific Discoveries of the Century Reveal God*, 2nd ed. (Colorado Springs, CO: NavPress, 1995), 138–144.
6. Ross, *Creator and the Cosmos*, 188–198; Hugh Ross, Kenneth Samples, and Mark Clark, *Lights in the Sky and Little Green Men: A Rational Christian Look at UFOs and Extraterrestrials* (Colorado Springs, CO: NavPress, 2002), 171–189.
7. "Fine-Tuning for Life on Earth," "Probabilities for Life on Earth," and "Fine-Tuning for Life in the Universe," http://designevidences.org, accessed April 2, 2006.

8. "Fine-Tuning for Life on Earth," "Probabilities for Life on Earth," and "Fine-Tuning for Life in the Universe," http://designevidences.org, accessed April 2, 2006.
9. Ross, *Creator and the Cosmos*, 175–199.
10. Ross, Samples, and Clark, *Lights in the Sky*, 171–189.

Chapter 11: Testing Creation/Evolution Models with Predictions
1. 1 Corinthians 15:14-15.
2. Hugh Ross, Kenneth Samples, and Mark Clark, *Lights in the Sky and Little Green Men: A Rational Christian Look at UFOs and Extraterrestrials* (Colorado Springs, CO: NavPress, 2002), 33–64; Fazale Rana and Hugh Ross, *Origins of Life: Biblical and Evolutionary Models Face Off* (Colorado Springs, CO: NavPress, 2004), 202–208.
3. D. Russell Humphreys, "Our Galaxy Is the Center of the Universe, 'Quantized' Red Shifts Show," *Creation Ex Nihilo Technical Journal* 18, no. 2 (2002): 1–10; R. V. Gentry, *Creation's Tiny Mystery*, 3rd ed. (Knoxville, TN: Earth Science Associates, 1992), 287–290; Jonathan Sarfati, *Refuting Compromise: A Biblical and Scientific Refutation of "Progressive Creationism" (Billions of Years) as Popularized by Astronomer Hugh Ross* (Green Forest, AR: Master Books, 2004), 156.

Chapter 12: Speeding the Truth Quest
1. The term "Christophobia" was first used by the Jewish legal scholar J. H. H. Weiler in his book *Un'Europa cristiana: Un saggio esploratiro* (Milan: Biblioteca Universale Rizzoli, 2003) and more broadly popularized by the Catholic theologian George Weigel in his book *The Cube and the Cathedral* (New York: Basic Books, 2005).
2. P. C. W. Davies and Charles H. Lineweaver, "Finding a Second Sample of Life on Earth," *Astrobiology* 5 (2005): 154–163.
3. Geochemical evidence establishes that life has been abundantly present on Earth as far back as 3.8 billion years (see chapter 6, pages 116–121). Metamorphic processes, though, have destroyed all fossils older than 3.5 billion years. Between 3.8 and 3.5 billion years ago, however, intense meteoritic and asteroidal bombardment of Earth resulted in the transport and deposit of over 10,000 tons of Earth material on every 100 square kilometers of the moon. See John C. Armstrong, Llyd E. Wells, and Guillermo Gonzalez, "Rummaging Through Earth's Attic for Remains of Ancient Life," *Icarus* 160 (2002): 183–196. Because the moon has lacked the metamorphic and tectonic forces that were prevalent on Earth, scientists possess an excellent opportunity of recovering on the moon pristine fossils of Earth's first life.
4. Eugenie C. Scott, "My Favorite Pseudoscience," *Reports of the National Center for Science Education* 23, no. 1 (2003): 11.
5. Lawrence M. Krauss, "'Creationism' Discussion Belongs in Religion Class," *Reports of National Center for Science Education* 22, nos. 1–2 (2002): 11.
6. Scott, "My Favorite Pseudoscience," 12.
7. Eugenie C. Scott, "The Big Tent and the Camel's Nose," *Reports of the National Center for Science Education* 21, nos. 1–2 (2001): 39.
8. Ken Ham, "Billions, Millions, or Thousands—Does It Matter?" *Back to Genesis*, no. 29 (May 1991): b.
9. As quoted by Skip Evans, "Ohio: The Next Kansas," *Reports of the National Center for Science Education* 22, nos. 1–2 (2002): 4.
10. A detailed account of this development may be found in Fazale Rana and Hugh Ross, *Origins of Life: Biblical and Evolutionary Models Face Off* (Colorado Springs,

CO: NavPress, 2004).
11. A detailed account of this development may be found in Hugh Ross, *A Matter of Days: Resolving a Creation Controversy* (Colorado Springs, CO: NavPress, 2004).
12. National Science Board, *Science and Engineering Indicators—2004*, vol. 2, Appendix Table 2-34, http://www.nsf.gov/statistics/seind04/append/c2/at02-34.pdf, accessed February 24, 2006.
13. National Science Board, *Science and Engineering Indicators—2004*.
14. Richard E. Smalley in a PowerPoint presentation, "Nanotechnology, the S&T Workforce, Energy, and Prosperity," to the President's Council of Advisors on Science and Technology (PCAST), March 3, 2003, slide 8, http://www.ostp.gov/PCAST/PCAST%203-3-03%20R%20Smalley%20Slides.pdf, accessed April 2, 2006.
15. Paul Chien, University of San Francisco biologist, in a recorded interview. Dr. Chien has made several visits to the Chengjiang shale and has published collaborative research with China's leading researchers. For an abridged version of the interview, see "Exploding with Life!" *Facts for Faith*, no. 9 (Q2 2000), 12–17.
16. Editors, "Responding to Uncertainty," *Nature* 437 (2005): 1.

Appendix A: Creation/Evolution Verbal Warfare
1. Henry M. Morris, "Is Creationism Important in Education?" http://www.answersingenesis.org/creation/v10/i3/education.asp, accessed July 29, 2005. Originally published in *Creation* 10, no. 3 (June 1988): 29–31.
2. Henry Morris, "Why ICR—and Why Now?" Impact, no. 337 (July 2001): ii, http://www.icr.org/index.php?module=articles&action=view&ID=448, accessed July 29, 2005.
3. John MacArthur, *The Battle for the Beginning: The Bible on Creation and the Fall of Adam* (Nashville, TN: W Publishing, 2001), 35.
4. Jason Lisle, "Feedback: Science Facts? Or Science Fiction?" http://www.answersingenesis.org/Home/Area/feedback/2005/0225.asp, accessed September 23, 2005.
5. Intelligent Design Network, Inc., *Seeking Objectivity in Origins Science* (Shawnee Mission, KS: Intelligent Design Network, 2002), 2.
6. Intelligent Design Network, 2.
7. Intelligent Design Network, 2.
8. Tim Stafford, "The Making of a Revolution," *Christianity Today*, December 8, 1997, http://arn.org/johnson/revolution.htm, accessed January 11, 2004.
9. Phillip Johnson, "The Church of Darwin," *Wall Street Journal*, August 16, 1999, http://arn.org/docs/johnson/chofdarwin.htm, accessed January 11, 2004.
10. Johnson, "Church of Darwin."
11. Michael Ruse, *Darwinism Defended: A Guide to the Evolution Controversies* (Reading, MA: Addison-Wesley, 1982), 303.
12. Stephen Jay Gould, *The Structure of Evolutionary Theory* (Cambridge, MA: Belknap Press, Harvard University Press, 2002), 982.
13. Niles Eldredge, *The Triumph of Evolution and the Failure of Creationism* (New York: Freeman, 2000), 149.
14. Ian Plimer, *Telling Lies for God: Reason vs. Creationism* (Milsons Point, New South Wales, Australia: Random House Australia, 1994), 1, 5.
15. Eugenie C. Scott, "Intelligent Design: Not Ready for Prime Time," Abstracts from the 22nd Annual Society for Scientific Exploration Meeting, June 8–11, 2003, in

Kalispell, Montana, in *The Explorer: Newsletter of the Society for Scientific Exploration* 18, no. 3 (2003): 5.

16. Lawrence S. Lerner, "The West Virginia Science Standards and Their Critics," *Reports of the National Center for Science Education* 23, no. 2 (2003): 5.

17. Jerry A. Coyne, "Creationism by Stealth," *Nature* 410 (2001): 745–746.

18. Mark Perakh, "A Presentation Without Arguments: How William Dembski Defeats Skepticism, or Does He?" http://www.talkdesign.org/faqs/present_arguments.html, accessed January 9, 2004.

Appendix B: Does the Constitution Bar Creation Teaching?

1. Martin Enserink, "Is Holland Becoming the Kansas of Europe?" *Science* 308 (2005): 1394; "Dealing with Design," *Nature* 434 (2005): 1053; Geoff Brumfiel, "Who Has Designs on Your Students' Minds?" *Nature* 434 (2005): 1052–1055; Geoff Brumfiel, "Biologists Snub 'Kangaroo Court' for Darwin," *Nature* 434 (2005): 550.

2. Robert T. Pennock, "Why Creationism Should Not Be Taught in the Public Schools," in *Intelligent Design Creationism and Its Critics,* ed. Robert T. Pennock (Cambridge, MA: MIT Press, 2001), 764.

3. Americans United for Separation of Church and State, http://www.au.org; People for the American Way, http://www.pfaw.org, accessed March 5, 2004.

4. *Epperson v. Arkansas*, 393 U.S. 97 (1968), 1.

5. *Epperson v. Arkansas*, 107–109.

6. *Epperson v. Arkansas*, 107.

7. Examples would be the state's intolerance of geocentrism and of the flat-earth hypothesis. Belief in geocentrism (the view that the earth is the center of the solar system and the universe) as a fundamental Christian doctrine is still held today by a few Christian organizations and churches (see note 36 of Appendix B). Well into the latter half of the 20th century the Flat Earth Society appealed to the Bible as an authority for its flat-earth doctrine (see note 35 of Appendix B).

8. Thomas Jefferson, "Letter of January 1, 1801," in *Thomas Jefferson: Writings*, ed. Merrill D. Peterson (New York: Library of America, 1984), 510.

9. The four references are (1) to "Nature's God" who is responsible for "the Laws of Nature," (2) "all Men are created equal . . . by their Creator," who endows them "with certain unalienable Rights," (3) "appealing to the Supreme Judge of the World," and (4) "firm Reliance on the Protection of divine Providence." For the full context, see John Hancock et al., "Declaration of Independence, in Congress, July 4, 1776: A Declaration by the representatives of the United States of America in General Congress assembled," in *The World Almanac and Book of Facts 1982*, ed. Hana Umlauf Lane (New York: Newspaper Enterprise Association, 1981), 466–467.

10. George Washington, "A Proclamation of National Thanksgiving, City of New York, October 3, 1789," in *On Faith and Free Government*, ed. Daniel C. Palm (New York: Rowman and Littlefield, 1997), 182.

11. U.S. Senate, "Chaplain's Office," http://www.senate.gov/reference/office/chaplain.htm, accessed March 4, 2004.

12. General Assembly of the State of Arkansas, "Act 590 of 1981, General Acts, 73rd General Assembly, State of Arkansas," in *Philosophy of Biology*, ed. Michael Ruse (Amherst, NY: Prometheus Books, 1998), 324.

13. *McClean v. Arkansas Board of Education*, 529 F.Supp 1255 (E.D. Ark 1982), section III, http://www.talkorigins.org/faqs/mclean-v-arkansas.html, accessed December 3, 2004.

14. *McClean v. Arkansas Board of Education*, sections IV(A), IV(D).
15. Testimony of Dr. G. Brent Dalrymple, *McLean v. Arkansas* Documentation Project, 410, http://www.antievolution.org/projects/mclean/new_site/pf_trans/mva_tt_p _dalrymple.html, accessed August 26, 2003.
16. Testimony of Dr. G. Brent Dalrymple, 411.
17. Eugenie C. Scott, *Evolution vs. Creationism: An Introduction* (Westport, CT: Greenwood Press, 2004), 108; Michael Ruse, *But Is It Science? The Philosophical Question in the Creation/Evolution Controversy* (Amherst, NY: Prometheus Books, 1996), 28; Edward J. Larson, *Trial and Error: The American Controversy over Creation and Evolution*, 3rd ed. (New York: Oxford University Press, 2003), 162–163.
18. Chandra Wickramasinghe, *Evidence in the Trial at Arkansas, December 1981 — What's New*, http://www.panspermia.org/chandra.htm, accessed March 10, 2004.
19. *McClean v. Arkansas Board of Education*, section IV(D).
20. *McClean v. Arkansas Board of Education*, section IV(D).
21. *McClean v. Arkansas Board of Education*, section IV(D).
22. *McClean v. Arkansas Board of Education*, 2.
23. *McClean v. Arkansas Board of Education*, 15, 19–20.
24. *Edwards v. Aguillard*, 482 U.S. 578 (1987), section I, http://caselaw.lp.findlaw.com/ scripts/getcase.pl?court=US&vol=482&invol=578, accessed February 23, 2005; Michael Shermer, *Why People Believe Weird Things: Pseudoscience, Superstition, and Other Confusions of Our Time* (New York: Freeman, 1997), 161–162.
25. *Edwards v. Aguillard*, statement 1(a) under "Held."
26. *Edwards v. Aguillard*, section Held: 1(a).
27. *Edwards v. Aguillard*, section III, A.
28. *Edwards v. Aguillard*, section Held: 1(a).
29. *Edwards v. Aguillard*, section Held: 2.
30. *Edwards v. Aguillard*, section Held: 2.
31. *Edwards v. Aguillard*, section Held: 1.
32. *Edwards v. Aguillard*, section III, A.
33. *Edwards v. Aguillard*, section Held: 1(b).
34. This point is conceded even by the nontheistic skeptic and anticreationist Michael Shermer in his book *Why People Believe Weird Things: Pseudoscience, Superstition, and Other Confusions of Our Time* (New York: Freeman, 1997), 162.
35. No major Christian denomination, either pre-Christopher Columbus or post-Christopher Columbus, has ever held to a flat-earth doctrine. However, several churches and a few small sects did teach that Earth is flat. Perhaps the most famous flat-earth proponent was the radio preacher Wilbur Glenn Voliva of Zion, Illinois. For a list of flat-earth advocates see http://www.loc.gov/rr/scitech/SciRefGuides/ flatearth.html, accessed April 4, 2004. The Flat Earth Society for its part has always claimed that the Bible is a flat-earth book. I heard this claim myself in the early 1990s when I participated in a radio debate on the shape of Earth on John Stewart's *Live from LA* program on KKLA, Los Angeles.
36. Prominent Protestant apologists for geocentrism include Gerardus Bouw, Malcolm Bowden, James Hanson, and Tom Willis. Leading Catholic apologists for geocentrism are R. G. Elmendorf, Paul Ellwanger, Paula Haigh, and Robert Sungenis. The two largest nonprofit organizations dedicated to promoting geocentrism are Catholic Apologetics International and The Biblical Astronomer. The following Web sites are devoted to promoting geocentrism: http://www.geocentricity.com, http://

www.fixedearth.com, and http://www.catholicintl.com, accessed April 4, 2004.

37. Hugh Ross, *The Creator and the Cosmos: How the Greatest Scientific Discoveries of the Century Reveal God*, 3rd ed. (Colorado Springs, CO: NavPress, 2001), 99–108. This reference includes all the citations for the original research sources.

38. Ross, 87–98, 169–174. This reference includes all the citations for the original research sources.

Appendix C: Biblical Origins of the Scientific Method

1. I establish in more detail the biblical origin for the scientific method in my book *The Genesis Question: Scientific Advances and the Accuracy of Genesis*, 2nd ed. (Colorado Springs, CO: NavPress, 2001), 195–197.

2. 1 Thessalonians 5:21.

3. Romans 12:2.

4. R. Laird Harris, Gleason L. Archer Jr., and Bruce K. Waltke, eds., *Theological Wordbook of the Old Testament* (Chicago: Moody, 1980), 51–52.

5. Joseph H. Thayer, *Thayer's Greek-English Lexicon of the New Testament* (Grand Rapids, MI: Baker, 1977), 512–513.

6. James 2:18.

7. Thomas F. Torrance, *Theology in Reconstruction* (Grand Rapids, MI: Eerdmans, 1965); Thomas F. Torrance, *Reality and Scientific Theology* (Edinburgh, UK: Scottish Academic Press, 1985); Thomas F. Torrance, "Ultimate and Penultimate Beliefs in Science," in *Facts of Faith & Science*, vol. 1, *Historiography and Modes of Interaction*, ed. Jitse M. van der Meer (New York: University Press of America, 1996), 151–176.

Appendix D: Functional Roles of "Junk" DNA

1. Sergei A. Konesev et al., "Neuronal Expression of Neural Nitric Oxide Synthase (nNOS) Protein Is Suppressed by an Antisense RNA Transcribed from an NOS Pseudogene," *The Journal of Neuroscience* 19 (1999): 7711–7720; Shinji Hirotsune et al., "An Expressed Pseudogene Regulates the Messenger-RNA Stability of Its Homologous Coding Gene," *Nature* 423 (2003): 91–96; Jeannie T. Lee, "Complicity of Gene and Pseudogene," *Nature* 423 (2003): 26–28; Evgeniy S. Balakirev and Francisco J. Ayala, "Pseudogenes: Are They 'Junk' or Functional DNA?" *Annual Reviews of Genetics* 37 (2003): 123–151.

2. Esther Betra'n et al., "Evolution of the *Phosphogycerate mutase* Processed Gene in Human and Chimpanzee Revealing the Origin of a New Primate Gene," *Molecular Biology and Evolution* 19 (2002): 654–663.

3. Christopher B. Marshall et al., "Hyperactive Antifreeze Protein in a Fish," *Nature* 429 (2004): 153.

4. Wen-Man Liu et al., "Cell Stress and Translational Inhibitors Transiently Increase the Abundance of Mammalian SINE Transcripts," *Nucleic Acid Research* 23 (1995): 1758–1765; Tzu-Huey Li et al., "Physiological Stresses Increase Mouse Short Interspersed Element (SINE) RNA Expression in *vivo*," *Gene* 239 (1999): 367–372; Richard H. Kimura et al., "Silk Worm Bm2 SINE RNA Increases Following Cellular Insults," *Nucleic Acid Research* 27 (1999): 3380–3387; Wen-Ming Chu et al., "Potential Alu Function: Regulation of the Activity of Double-Stranded RNA-Activated Kinase PKR," *Molecular and Cellular Biology* 18 (1998): 58–68.

5. Wen-Man Liu et al., "Alu Transcripts: Cytoplasmic Localisaton and Regulation by DNA Methylation," *Nucleic Acid Research* 22 (1994): 1087–1095; Wen-Man Liu et

al., "Proposed Roles for DNA Methylation in Alu Transcriptional Repression and Mutational Inactivation," *Nucleic Acid Research* 21 (1993): 1351–1359; Carol M. Rubin et al., "Alu Repeated DNAs Are Differentially Methylated in Primate Germ Cells," *Nucleic Acid Research* 22 (1994): 5121–5127; Igor N. Chesnokov and Carl W. Schmid, "Specific Alu Binding Protein from Human Sperm Chromatin Prevents DNA Methylation," *Journal of Biological Chemistry* 270 (1995): 18539–18542; Utha Hellman-Blumberg et al., "Developmental Differences in Methylation of Human Alu Repeats," *Molecular and Cellular Biology* 13 (1993): 4523–4530.

6. Jeffrey A. Bailey et al., "Molecular Evidence for a Relationship Between LINE-1 Elements and X Chromosome Inactivation: The Lyon Repeat Hypothesis," *Proceedings of the National Academy of Sciences, USA* 97 (2000): 6634–6639; Mary F. Lyon, "LINE-1 Elements and X Chromosome Inactivation: A Function for 'Junk' DNA?" *Proceedings of the National Academy of Sciences, USA* 97 (2000): 6248–6249.

7. Edith Heard et al., "X Chromosome Inactivation in Mammals," *Annual Review of Genetics* 31 (1997): 571–610; Jack J. Pasternuk, *An Introduction to Human Molecular Genetics: Mechanisms of Inherited Diseases* (Bethesda, MD: Fitzgerald Science Press, 1999), 31–32.

8. Elena Allen et al., "High Concentrations of Long Interspersed Nuclear Element Sequence Distinguish Monoallelically Expressed Genes," *Proceedings of the National Academy of Sciences, USA* 100 (2003): 9940–9945.

9. Alam G. Atherly, Jack R. Girton, and John F. McDonald, *The Science of Genetics* (Fort Worth, TX: Saunders College Publishing, 2000), 597–608; Greg Towers et al., "A Conserved Mechanism of Retrovirus Restriction in Mammals," *Proceedings of the National Academy of Sciences, USA* 97 (2000): 12295–12299; Jonathan P. Stoye, "An Intracellular Block to Primate Lentivirus Replication," *Proceedings of the National Academy of Sciences, USA* 99 (2002): 11549–11551; Theodora Hatziioannou et al., "Restriction of Multiple Divergent Retroviruses by LV1 and Ref1," *The European Molecular Biology Organization Journal* 22 (2000): 385–394.

10. François Mallet et al., "The Endogenous Retroviral Locus ERVWE1 Is a Bona Fide Gene Involved in Hominoid Placental Physiology," *Proceedings of the National Academy of Sciences, USA* 101 (2004): 1731–1736.

11. Clare Lynch and Michael Tristem, "A Co-opted Gypsy-Type LTR-Retrotransposon Is Conserved in the Genomes of Humans, Sheep, Mice, and Rats," *Current Biology* 13 (2003): 1518–1523.

12. Vera Schranke and Robin Allshire, "Hairpin RNAs and Retrotransposon LTRs Effect RNAi and Chromatin-Based Gene Silencing," *Science* 301 (2003): 1069–1074; Wenhu Pi et al., "The LTR Enhancer of ERV-9 Human Endogenous Retrovirus Is Active in Oocytes and Progenitor Cells in Transgenic Zebrafish and Humans," *Proceedings of the National Academy of Sciences, USA* 101 (2004): 805–810; Catherine A. Dunn, Patrick Medstrand, and Dixie L. Mager, "An Endogenous Retroviral Long Terminal Repeat Is the Dominant Promoter for Human b1,3-Galactosytransferase 5 in the Colon," *Proceedings of the National Academy of Sciences, USA* 100 (2003): 12841–12846.

Appendix E: the Purpose and Extent of Noah's Flood

1. Genesis 7:19; 8:5-9; Job 38:8-11; Psalms 33:6-7; 104:5-9; Proverbs 8:22-30; Ecclesiastes 1:4-7; 2 Peter 3:4-6.

2. Referring to the Genesis 1:9-10 account.

3. Job 38:8-11; Proverbs 8:24-29.

4. 2 Peter 3:6.

5. Joseph Henry Thayer, *Thayer's Greek-English Lexicon of the New Testament* (Grand Rapids, MI: Baker, 1977), 629.

6. Genesis 1:28; 9:1,7.

7. Genesis 11:8-9.

8. Fazale Rana with Hugh Ross, *Who Was Adam? A Creation Model Approach to the Origin of Man* (Colorado Springs, CO: NavPress, 2004), 67.

INDEX

ABOUT HUGH ROSS

Hugh Ross is founder and president of Reasons To Believe. As a boy, he studied the stars using a homemade telescope built with the proceeds from collected beverage bottles. As an astronomer, he relies on more advanced instruments to gaze into the depths of space and time—back to the earliest detectable moments of the cosmos.

With grants from the National Research Council of Canada, he earned a BS in physics from the University of British Columbia and later an MS and a PhD in astronomy from the University of Toronto. For several years he continued his research on quasars and galaxies as a postdoctoral fellow at the California Institute of Technology. During that time he began more than two decades of service on the pastoral staff at Sierra Madre Congregational Church.

Today, in addition to overseeing the operations of Reasons To Believe and hosting a weekly live Webcast, *Creation Update*, Dr. Ross speaks on university campuses and in a variety of venues around the world. His books include *The Fingerprint of God*, 2nd ed. (Whitaker House, 2000), *Beyond the Cosmos*, 2nd ed. (NavPress, 1999), *The Creator and the Cosmos*, 3rd ed. (NavPress, 2001), *The Genesis Question*, 2nd ed. (NavPress, 2001), *Lights in the Sky and Little Green Men* (with Kenneth Samples and Mark Clark, NavPress, 2002), *Origins of Life* (with Fazale Rana, NavPress, 2004), *A Matter of Days* (NavPress, 2004), and *Who Was Adam?* (with Fazale Rana, NavPress, 2005). He has contributed to many other works, including most recently *What God Knows* (Baylor University Press, 2005).

Dr. Ross lives in Southern California with his wife, Kathy, two sons, two cats, and one dog.

ABOUT REASONS TO BELIEVE

Founded in 1986, Reasons To Believe (RTB) is a nonprofit organization, without denominational affiliation, adhering to the historic Christian creeds and affirming both the accuracy and authority of Scripture. RTB exists to research and communicate the compatibility of God's revelation in the words of the Bible with the established facts of nature, including the latest findings of scientific research.

RTB staff and volunteer speakers are available to address schools (elementary to graduate level), churches, business firms, and community groups. *Creation Update*, a Webcast reporting on recent discoveries and their relevance to the Christian faith, airs (live) each Tuesday from 11 a.m. to 1 p.m. (PST) on www.reasons.org (where programs and notes are also archived) and on www.oneplace.com. A hotline for people with questions about science, faith, and the Bible operates daily from 5 to 7 p.m. (PST) at (626) 335-5282. Science apologetics courses (for university credit or audit) are available on the Internet through Reasons Institute. See RTB's Web site for details.

For further information about RTB or to receive a free newsletter and catalog of resources, contact the ministry by phone at (800) 482-7836, by mail at P.O. Box 5978, Pasadena, CA 91117, or on the Web at www.reasons.org.

DISCOVER MORE REASONS TO BELIEVE.

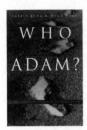
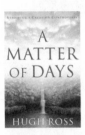
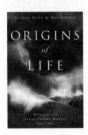